Economics of Forestry and Rural Development

Economics of Forestry and Rural Development

An Empirical Introduction from Asia

*William F. Hyde, Gregory S. Amacher,
and Colleagues*

Ann Arbor
THE UNIVERSITY OF MICHIGAN PRESS

Copyright © by the University of Michigan 2000
All rights reserved
Published in the United States of America by
The University of Michigan Press
Manufactured in the United States of America
♾ Printed on acid-free paper

2003   2002   2001   2000   4 3 2 1

No part of this publication may be reproduced,
stored in a retrieval system, or transmitted in any form
or by any means, electronic, mechanical, or otherwise,
without the written permission of the publisher.

A CIP catalog record for this book is available from the British Library.

Library of Congress Cataloging-in-Publication Data

Economics of forestry and rural development : an empirical introduction from Asia / [edited by] William F. Hyde, Gregory S. Amacher, and colleagues.
    p. cm.
  Includes bibliographical references.
  ISBN 0-472-11144-2 (cloth : alk. paper)
    1. Forests and forestry —Economic aspects—Asia.   2. Rural development—Asia.   I. Hyde, William F.   II. Amacher, Gregory S. 1962-

SD219.E36 2000
333.75′15′095—dc21   00-062999

Contents

Preface vii

Part 1. Introduction

1. A General Statement: Nineteen Hypotheses about Forestry and Rural Development 3
   William F. Hyde and Gregory S. Amacher

Part 2. Background

2. Forest Environments as Attractants for Human Migration: The Philippines in the 1980s 29
   Gregory S. Amacher, Ma. Concepcion J. Cruz, Wilfrido Cruz, and William F. Hyde

3. Some General Features of Household Forestry in Nepal 43
   K. H. Gautam, B. R. Joshee, R. L. Shrestha, V. K. Silwal, M. P. Suvedi, and L. P. Uprety with William F. Hyde

Part 3. Household Production and Consumption, and the Adoption of New Technologies

4. Household Fuel Production and Consumption, Substitution, and Innovation in Two Districts of Nepal 57
   Bharat R. Joshee, Gregory S. Amacher, and William F. Hyde

5. Innovation and Adoption in Pakistan's Northwest Frontier Province 87
   Mohammad Rafiq, Gregory S. Amacher, and William F. Hyde

Part 4. Regional Supply and Demand

6. Estimates of Economic Supply from Physical Measures of the Forest Stock: An Example from Eight Developing Countries 103
   Kerry Krutilla, Jintao Xu, Douglas F. Barnes, and William F. Hyde

7. Regional Fuelwood Production and Consumption in Nepal: With Implications for Local Adoption of New Forestry Practices 121
   Keshav R. Kanel, Gregory S. Amacher, William F. Hyde, and Lire Ersado

Part 5. Secure Rights

8. Secure Forest Tenure, Community Management, and
   Deforestation: A Philippine Policy Application   151
   *Marcelino Dalmacio, Ernesto S. Guiang, Bruce Harker,
   and William F. Hyde*

9. Rural Reform and China's Forestry Sector: Rational Farmer
   Expectations and the Case for a Stable Policy Environment   181
   Runsheng Yin, David H. Newman, and William F. Hyde

Part 6. Additional Perspective

10. Common Property Resources and the Dynamics of Rural
    Poverty: Field Evidence from the Dry Regions of India   203
    N. S. Jodha

11. Trees as a Source of Agricultural Sustainability:
    Agroforestry in China   223
    Runsheng Yin and William F. Hyde

Part 7. Conclusions

12. Social Forestry Reconsidered   243
    William F. Hyde, Gunnar Kohlin, and Gregory S. Amacher

Preface

Forestry, in some form, has always been a component of rural life, and particularly the life of the lowest income citizens of isolated rural communities. Since the early 1970s, forestry and rural development, which in some forms is also known as "social forestry," has also been a source of hope for some and intellectual inquiry for others involved in the activities of international development.

When we began this book a decade ago the economic component was either weak or absent from discussions of forestry and rural development. So was the careful empirical evidence that could support economic intuition—or even opinion. Anthropologists, and perhaps other social scientists, had been up to the task but the practitioners of economics had not.

We saw our task as providing initial empirical economic reasoning for a deserving subject. In the course of ten years we have had a lot of assistance from many colleagues, some of whom became co-authors of chapters in this book. We hope we have accomplished our task. Also in the course of ten years many others have begun to address the same economic questions. The intellectual capital on social forestry or forestry and rural development is much greater than it was a decade ago. Our final chapter surveys the expanding literature and identifies some important contributions—and contributors—as well as some remaining unresolved issues.

We believe our book makes five important contributions:

- We have sharpened the economic definition of some questions about forestry and rural development. Forests are not "good," neither is their degradation "bad"—except in terms of an effect on social welfare. Foresters and economists have understood this point for years but their discussions often feature gross rather than net effects on welfare, and their focus has been on the forests of developed countries. Perhaps we have added clarity to the contributions of forests to net welfare, particularly for subsistence economies and especially in the developing countries of Asia.
- We have introduced the household economics literature to forestry (chapters 4, 5, and 7). The household analytic

approach provides a powerful and rigorous means for examining cases in which households are both producers and consumers of the same good. This condition characterizes many subsistence users of the forest in developing countries. We would suggest that it also characterizes many non-industrial private forest landowners in North America and Europe. The household approach should be useful in assessing the revealed preferences of these landowners as well. Subsistence resource users must be incorporated in the targets of policy decisions. Market consumption and production evidence alone disregard subsistence households and can only lead to egregious errors, especially in rural communities where subsistence activities comprise a large share of the total impacts on the forest.

- There is no shortage of useful data for economic analyses of developing country forestry (chapters 2-5, 7, 10). In fact, to us it often seems like every master's degree student from a developing country has the data to address another interesting social forestry question. On the other hand, we have demonstrated that the standard physical measures of forests are weak proxies for economic measures of the same resources (chapters 3, 4, 6-8, 10). Their uncritical use misleads many discussions of forest policy, and especially many current discussions of global deforestation.

- We have participated in the growing discussion of the impacts of "non-forest" policies on forests (chapters 2, 8, 11). Most researchers and policy analysts are only beginning to understand the effects of the broader policy environment on the forest, but we would venture an argument that the broader policy environment often dominates the effects of all forest policies combined. So, neither deforestation, nor local community uses of the forest can be understood without also understanding the effects of agricultural, macroeconomic, and trade policies on the forest. This can only increase the importance of accurate "targeting," the careful matching of the policy instrument with the policy objective.

- Finally, our summary insights come together in a universal pattern of forest development (chapter 12). This pattern shows that the truly unique features of forestry are its three margins of land use (extensive and intensive margins for plantations plus the margin of natural forest) rather than the usual two, and the efficient open access condition at the third margin. This book contains the fullest description of this pattern to data (chapters 8 and 12 together). We believe broader recognition of this simple pattern would prevent many policy and programmatic errors in forestry and in protecting forest environments.

We have our own faith in these contributions but we also recognize that the judgment of our readers is the critical test!

Let's consider those readers: Our intention throughout the book was to produce reliable economic assessments. This implies considerable technical detail in many chapters. For this reason, we tried to produce introductory and concluding chapters with sufficient contextual background to introduce the technical chapters. We also included introductory and concluding sections to each chapter that should further ease the reader's way through the more detailed technical material. We hope graduate students in forest economics and policy or in rural development will be able to read the entire manuscript without difficulty, and we hope others will be able to understand our organization and conclusions regardless of the underlying technical justifications.

Many have assisted and encouraged us through the long course of preparing this book. Bob Gregory of the University of Michigan was the first to introduce both of us to many questions of forestry and rural development. Bob, and especially Jeff Romm and the Ford Foundation that supported both of them, were pioneers in our field of inquiry. Bill Bentley and Dave Nygaard, through Winrock International, funded a sabbatical at Kasetsart University in Bangkok that got us started on the book itself. Bill's encouragement has been steadfast throughout. Kasetsart was the source of many good lifelong associations. Of course, our several chapter co-authors are some of those lifelong associates, and we owe them for very many of the insights and much of the quality of this book.

Friends like Ann delos Angeles (first the Philippine Institute of Development Studies, then REECS in Manila), David Griffin (then at ANU), Neil Byron (first at ANU in Canberra, then with CIFOR in Bogor), George Taylor (USAID), Larry Hamilton (then at the East-West Center), Josh Bishop (IIED in London), Frank Convery and Peter Clinch (University College/Dublin), Bruce Larson (then with the Economic Research Service in Washington), Willy and Chona Cruz (first at University of Philippines in Los Banos, now with the World Bank), Juan Seve (IRG, Inc in Washington, Manila, and Jakarta), and Bill Magrath (World Bank) each shared insights that helped us over major hurdles. In fact, Ann's own dissertation could claim to be midwife to many of our initial thoughts. Percy Sajise and Hermi Francisco (UP/Los Banos), Asa Sajise (Cal-Berkeley), Gershon Feder, Luis Constantino, and Ken Chomitz (all three with the World Bank), C. Thangamuthu

(Bharathidasan University), Aimo Juhola (first with Asia Development Bank, then with Jakko Poyry), Virgilio Viana (IMAFLORA and the University of Sao Paolo), Steve Stone (first Cornell, and now the InterAmerican Development Bank), Barin Ganguli (first ADB, then Jakko Poyry), Alemu Mekonnen (Addis Ababa University), Roger Sedjo (Resources for the Future), Arun Malik (George Washington University), Woon Chuen (FRIM in Malaysia), and Will Knowland (then with USAID) each shared a key insight or provided an important critique at a moment when we were particularly alert for it.

And finally, we owe untold debts to seminar participants at universities and research institutions on six continents on whom we tried out many of our arguments and incomplete analyses. Our debt is greatest to these at the University of Alberta's Department of Rural Economy and at Thomas Sterner's Environmental Economics Unit at Goteborg University who sat through several seminars, and who continue to share their own useful insights regardless of the repetition. We would like to compliment those two institutions for the meeting places they provide for many who share global interests in economic assessments of policy and the rural natural environment.

Dolly Tiongco and Darcy Amacher, Big John and Chase, have shared our enthusiastic moments, suffered our dejected moments, and supported us throughout unconditionally. Our wives and sons have given us seven great years in the process. We dedicate this book to the four of them.

>William F. Hyde
>Centre for International Forestry Research and
>Forest Economics and Policy Analysis Research
>Centre, University of British Columbia
>
>and
>
>Gregory S. Amacher
>Department of Forestry, Virginia Polytechnic
>Institute and State University

Part 1
Introduction

CHAPTER 1

A General Statement: Nineteen Hypotheses about Forestry and Rural Development

William F. Hyde and Gregory S. Amacher[1]

The successful pursuit of forestry for economic development once suggested large scale timber and fiber operations. More recently, we have come to understand that another variety of forestry contributes in important ways to the economic well-being of some of the world's poorest populations. This second variety, sometimes known as "social forestry," has to do with the local use of trees and forests for domestic consumption. These local household uses of forest resources are the topic of this book.

The local uses of forest resources is an exciting topic of both intellectual inquiry and public action for both foresters and those with more general interests in rural development. The poor, often subsistence, economies associated with forests attract our sensitivities for rural poverty and for social welfare in general. The marginal, often fragile, physical environments associated with forests attract our concerns for resource conservation. The attraction is all the greater because the topic extends beyond national, and even continental, boundaries. It includes local indigenous initiative as well as activities sponsored by domestic forestry and rural development agencies and by international donor agencies.

The term "social forestry" was originally associated with this experience as it was applied to the Indian subcontinent—but its use has expanded. Contemporary discussions of forestry and local rural development reach beyond the communal orientation of "social forestry" and they certainly extend beyond South Asia. Contemporary discussions include a spectrum of institutional arrangements for the rights to trees and forestlands, and for the substitutes that would reduce the local demands on trees and forests. While these discussions typically feature the developing countries of Africa and Asia, the more heavily forested regions of Latin America and even some farm forestry applications in the developed countries of North America and Northern Europe also find their way into the literature.

Forestry traditionally focused on commercial production of timber and fiber. Traditional training in forestry still features either forest protection or production of these two outputs. Commercial production and traditional training contrast with a focus on local rural development, however, in that they often compete with many of the forest products consumed in rural societies: i.e., fuelwood, fodder, forage, fruits, and other domestically consumed non-wood forest crops.

This diversity of products is an indication of the range of interests that attract our attention and also of the multiple of possible responses to local consumer demands. The diversity of human institutions, particularly the many customs and conventions for property rights in trees and in forest land, is another. Diversity is an indication of the adaptability of forest-related activities to a broad cross-section of development situations.

Forest products have always been important to indigenous human populations, but social forestry has only been a topic of serious inquiry by foresters and rural developers for, perhaps, the last thirty years. It is still in its earliest and most formative years. The term "social forestry" first appeared in Gujarat in the mid-1960s. Jack Westoby gave the term broader recognition in his address to the Ninth Commonwealth Forestry Conference in 1968. The Ford Foundation in the early 1970s—with the special insights of Jeff Romm, Marshall Robinson, and their Asian colleagues—provided further discussion and organized financial support. Other individual observers and other development agencies have extended these Gujarat, Westoby, and Ford observations and there is now a broad literature composed of many casual observations and a few informal hypotheses.[2] The next steps in the intellectual development of the topic require data analysis and the rigorous empirical and quantitative examinations of formally-stated hypotheses. Our objective in this volume is to begin these next steps.

We cannot be comprehensive. The topic is too broad, its geographic and climatic range too extensive, and the affected human populations and interests too diverse. Our alternative is to present a series of case studies which cut across a spectrum of regional and conceptual characteristics. Our region of inquiry is Asia—from the Philippines and China in the east through Indonesia, India, and Nepal to Pakistan in the west. This region—and our cases—includes tropical and temperate, sparse and dense, upland and lowland forests; arid and wet climates; and the full range of human population densities.[3]

Our case studies occur in sets. Most are local and specific but each set of cases intends to be sufficiently diverse to suggest generalizations. The first set furnishes examples of the broad and general importance of trees and the forest to local households and to potential immigrants from more distant populations as well. The second set more precisely examines the allocation of household resources to forest production and household consumption, as well as the demographic and social characteristics that explain the local acceptance of new social forestry activities.

The third set moves beyond the household and local levels of analysis and begins to inquire about regional demand and supply. Regional evidence of the relative importance of standing forests, the reliability of existing markets, and the opportunities for substitution will help identify the geographic targets for policies and technologies that can make a difference.

This would be a comprehensive collection of cases if land tenure were not such a critical issue. Forests and trees are dispersed and generally low-valued resources that tend to grow at the margin of economically productive land. The rights to these resources are often ill-defined. But these rights are critical for the poorest people and for sustainable management of the resource. The rights become more clearly defined with resource depletion, rising prices, and the passage of time. Our fourth set of cases examines these issues. We will find that establishing secure tenure is important but that its impact can be limited by exogenous factors such as the stability of national policies. The policy environment can have unintended spillover effects that destroy well-designed direct forestry sector incentives.

The final two cases raise two remaining and unsettled issues: the impacts of these forest-related development activities on poor and especially landless households, and their impacts on the environment. The former is unknown, while the latter seems settled in the favor of environmental improvement but little evidence aside from controlled research plots supports this contention. Finally, our book closes with a chapter that summarizes, contemplates generalizations arising from the previous chapters and from other recent literature, and suggests topics for further inquiry.

We will find that forestry's impact on local consumption in poor communities is both greater and different than often anticipated. For example, the availability of unclaimed forest land has attracted an upland migrant population that is eighteen times greater than the 1980 estimate of the Philippine Bureau of Forest Development. Fuelwood scarcity, however, may be less a problem in its own right—even in Nepal's hills where the standing forest inventory may be sparse and some farm households plant their own trees. Yet it may be more of a problem if it diverts increasing quantities of scarce household labor from essential agricultural and food preparation activities. Will the labor diversions be an increasing burden on women—as many suspect? Finally, we might argue that neither fuelwood scarcity nor the associated diversions of household and agricultural labor for fuelwood collection will become problems where rights to property like trees and forest land are clear and secure. On the other hand, we will argue that the transactions costs for establishing these rights will always exceed their value on some land and, therefore, some forests and forestry activities will never be sustainable.

We will find substance behind the idea that forestry activities can provide important basic support for carefully selected poor rural populations early in the development process. Numerous factors will affect the success of these activities. We will hypothesize that success often depends more importantly on a) economic scarcity, b) the opportunities for substitution, c) secure resource tenure, d) the rate of acceptance by local leaders who, while poor, are certainly are not the poorest of the poor, e) local respect for the forestry or development agency's advisors, and f) exogenous policies with unintended impacts on social forestry—than on some of the more usual concerns of forest policy and management.

Background and Definition

The remainder of this chapter is a more detailed introduction to social forestry. It begins with a definition of social forestry that suits the purpose of this book, and follows with a discussion of social forestry's broadest objectives. Local initiative is often the medium for accomplishing these objectives. Local initiative, however, often finds support from economic development projects funded by domestic resource management agencies or external donor. Therefore, this discussion continues with a brief review of reasonable expectations for such projects. These expectations are an introduction to the behavior of households and local markets observed in many of our case studies. Finally, local property rights and resource tenure, as well as the long-term reliability of these policies and also exogenous policies designed for altogether different target sectors, can frustrate the greatest financial incentives of the most carefully designed development project. Therefore, our background discussion finishes with an introduction to the topics of secure resource tenure and general economic policies that can unintentionally and indirectly affect social forestry in a substantial way. In sum, this background section intends to motivate our subsequent case studies.

Numerous terms are associated with forestry for local use in rural development and various definitions are associated with each of these terms.[4] The most common terms are "social forestry," "community forestry" and "agroforestry." Their definitions have been diverse and exclusive in order to satisfy the specific purposes of the many different users. Perhaps we can best satisfy our purposes with a comprehensive definition which features the exclusions. That is, "forestry for local use in rural development," which we abbreviate as "social forestry," is any forestry except large scale commercial plantations so long as it emphasizes the responses of local consumers to forest-based goods and services: usually fuelwood, fodder, and forage, sometimes water, soil protection, and other tree and interplanted non-wood crops.

Consider this definition carefully. It includes domestic consumption of household-produced forest products and it includes market exchange. It incorporates the original concept of social forestry as well as the concepts of community forestry and agroforestry. Social forestry has always referred to local use and rural development, but it also has a strong, and sometimes restrictive, association with South Asia. Community forestry usually refers to commonly owned or controlled forests and agroforestry refers to the farming systems involved in growing trees as a crop or growing intermixed trees and agricultural crops. In the context of this book, however, and from the perspective of opportunities for economic development, our use of the term social forestry extends past the Indian subcontinent to include, for example, the rural migration and new upland settlement and land tenure issues common to Laos, Thailand, and the Philippines, as well as any other contemporary forestry issue of importance to local people and rural development.[5]

Our view of social forestry is not restricted to subsistence economies and communal activities, and market exchange does not hinder our definition. Production for household consumption is a fact but very few, if any, poor farm families exist solely on their own domestic production. Most households offer some labor or agricultural products in the local market, receive some currency in exchange, and purchase some share of their total consumption with it. Therefore, some fuelwood and fodder production may be consumed by producer families but we should not be surprised, and we will not alter our definition of social forestry, if some also exchanges for currency in local markets.[6]

Neither local institutional distinctions nor the distinctions between community forestry and agroforestry affect our definition of social forestry. Both social forestry and local development suggest increasing market diversity and shifting incentives for resource management. New incentive structures in turn suggest changing institutions, particularly the institutions explaining local property right arrangements for the relevant resources: trees, land, fodder, and forage. Therefore, shifts from established common ownership and management arrangements to more individual and private property arrangements often accompany economic development. The task of designing successful forestry development programs assumes new difficulties as a result of these dynamic events. There are important opportunities for both community forestry and agroforestry but successful community forestry activities are more difficult to design than successful agroforestry activities because agroforestry most often involves small private landholdings where the incentives and the target population are clearer. Moreover, we might anticipate that the ambitious subset of all landowners who willingly accept the risks and who can afford the costs associated with initiating new forestry investments are more promptly responsive to market changes, while decision makers with more complex communal responsibilities require longer adjustment periods to organize communal responses. Altogether, this suggests integral roles for property rights and common access in any assessment of social forestry, as well as the categorization of both community forestry and agroforestry as special applications of our rural development interests.

The Objectives of Social Forestry

If our primary objective is to use forest resources to assist local community development, then we can probably say that the responsiveness of local consumers to any forest development activity largely defines its limits. The communities in question tend to be rural and poor. Therefore, helping the rural poor is a complementary objective of many social forestry activities. The first objective is an efficiency objective and the second is an income distributive objective. We must judge the quality of any social forestry activity by its success in achieving these two objectives. Therefore, our case studies must provide insight to the design and location of activities which would satisfy these two basic objectives.

Efficiency implies a concern for economic growth with, in this case, social forestry as the means. It means that the marginal social benefits exceed the marginal social costs of acceptable activities, and that they promote economic growth. For example, public participation and improved seedlings are useful inputs and halting deforestation and controlling erosion are useful outputs only if, in the final social account, the foregone marginal resource opportunities associated with these and other inputs are less than the marginal gains associated with the resulting outputs. For example, we can justify diverting land and labor from agriculture to forestry if the value of the new forest products is greater that the value of the foregone agricultural opportunity.

Efficiency means that the inputs and outputs of social forestry or of any other physical resource project are justified only by satisfying this rigorous test. Locally initiated activities necessarily pass this test--or else the local participants would discontinue the activity. The test must be more formal for resource management ministries and international donor agencies who have greater resources, less local contact, and greater opportunity to absorb and overlook their own local failures.

The distributive objective invites more extensive discussion beginning with two questions and continuing with some thought about the diffusion of new ideas among poor populations. First, do we care whether the benefits of social forestry activities reach all the poor or is it acceptable if they reach only some groups of the poor? Must some benefits accrue to poorest of the poor? Must they accrue immediately? Second, do we care whether beneficial activities help the not-so-poor so long as they also help the poor? What if the not-so-poor gain more than the poor but the poor gain nonetheless?

Distributive questions like these are one entry point for the important discussions about women's roles in forestry. We can substitute "women" for "poor" and ask the same questions. Indeed, women often are the poorest of the poor—although it clearly would be a mistake to use the terms "women" and "poor" interchangeably.

The answers to these distributive questions are subjective. They may vary with one's personal perspective and they certainly vary with the geographic and demographic boundaries of inquiry. They are important because rarely does any activity, social forestry or otherwise, benefit all of any target population and rarely is any target group of benefactors universally poor. Indeed, all too often the greatest benefactors are not the truly poor, despite the stated program objective. Social forestry activities, for example, may intend to benefit the very poor but they often have their most immediate impacts on aggressive and receptive larger landowners. Is this a bad outcome if even the (recipient) largest landowners are poor from a policy maker's or a donor agency's perspective and if small (unaffected) landowners and the landless are just poorer yet? Is it a bad outcome if large landowners absorb the risks of trying a new activity and also accumulate the first benefits, but small landowners and the landless follow the demonstrations of their better-off neighbors to obtain subsequent benefits for themselves?

Clearly the assertion that a program intends distributive gains is insufficient. Thoughtful analysis is a prerequisite to understanding whether the target population is a realistic benefactor of the action in question and, if realistic, whether this population is truly deserving of redistributive gain.[7]

Household Budgets and Consumption Elasticities

These distributive issues cause us to ponder whether there are any generalizable expectations which might guide programs and policies designed with distributive objectives. Perhaps there are two: search for goods and services that require significant household budget shares (where "budget" should be interpreted broadly to include all household resources, financial and otherwise) and for which the targeted population displays large income elasticities and a willingness to substitute.

The advantage of the former should be obvious. If the budget share is small, then the distributive impact of the best-conceived program may be desirable but it can only be small. The likelihood of a meaningful impact is greater where the budget share is greater.

It is unclear what budget shares the rural poor spend on their forest products. Indeed, this is an important empirical question and the answer probably varies widely by products and across regions and cultures. Nevertheless, it is a widely held expectation that the expenditures for fuelwood, fodder and forage are proportionately greater for the very poorest households. For example, most poor people burn wood for both heating and cooking. Those who are too poor to afford fuelwood may still prefer it to the alternatives of dung or combustible agricultural residues. Fuel is a primary consumption good in the poorest households. Therefore, it may exhaust a significant share of the cash budget or, especially, the budget of available labor in the poorest households. This expectation would encourage many social forestry activities.

The elasticity criterion refers to an increase in consumption relative to an increase in income, over some income range. A large income elasticity implies that the distributive objective probably remains valid even as the region begins to move along a path of economic development and the target population's income and wealth begin to rise. It also implies that the distributive objective may be satisfied for households with widely disparate wealth levels.

Income elasticity provides only a partial indication of a household's willingness to substitute the products of a new social forestry activity for its current consumer goods. Understanding substitution also requires insight to the household's responsiveness to changes in the scarcities of forest products and their substitutes. Larger price and substitution elasticities indicate that the household already perceives scarcity and is changing its behavior in response. Small price and substitution elasticities indicate either a lack of substitution opportunity or the household's demonstrations that increasing scarcity for the good in questions is not important enough to induce a change in household behavior.

For example, consider a region where fuel consumes substantial shares of household income and labor opportunity, the income elasticity of fuelwood is close to one, the fuelwood price elasticity is also nearly unitary, and the substitution elasticity between fuelwood and combustible agricultural residues is large. This region would be a good candidate for an agroforestry project that introduced new fuelwood sources (like seedlings) or technical substitutes (like improved stoves) with the objective of reducing the consumption of residues (and enhancing long-term soil productivity). The opposite conditions, on the other hand, would indicate that success of new agroforestry activities is less likely. We will empirically examine fuelwood budgets, and income, price, and substitution elasticities in various of our case studies from Nepal, Pakistan, and India.

Investment Patterns and Technology Transfer

Our knowledge of diffusion patterns for new technologies in other populations suggests a second generalizable expectation. The aggressive, educated and wealthier members of any society are usually the first to accept new ideas and to adopt new techniques (Feder et al. 1985, Rogers 1983, Ruttan 1977, Schultz 1964). Certainly it would be unusual for the poorest members who have the least amount of risk capital (discretionary land, household labor, or financial capital) to be the first to try a new idea—including improved seedlings, new planting or management strategies, new marketing opportunities, improved stoves, or other new social forestry techniques.

One implication of this observation is that well-designed projects must include local demonstrations in order to achieve a satisfying rate of acceptance. Furthermore, we must probably accept that the initial local respondents to any new technology will be among the better-off community members who, perhaps, are not among the target population for the eventual distribution of benefits. These wealthier respondents can bear the initial risks and their successes will demonstrate the merit of new techniques to those poorer households who may comprise the real target population. Poor households often cannot afford the risk of failure of even a single crop on a single small garden plot. Therefore, they cannot afford the personal uncertainty accompanying innovation. They may rapidly adopt proven techniques, however, after their wealthier neighbors demonstrate some initial success. Our second set of case studies provides evidence of local economic and demographic characteristics explaining the successful adoption of social forestry activities in Nepal and Pakistan.

These observations on diffusion have the potential to be misunderstood as a "trickle down" justification for what is actually redistribution to the wealthy. We must carefully guard against this. Where the improved well-being of the rural poor is an important component of social forestry, then legitimate diffusion arguments must favor that subset of all programs and policies which would have favorable initial affects on those who are better off, but which have clear and substantial positive secondary impacts on their poorer neighbors.

Fortunately, policy makers and project planners can restrict perverse redistributive policies by insuring that they couple an understanding of diffusion with the prior rule for outputs associated with large household budget shares, larger income elasticities, and elastic prices for the target population. This coupling restricts us from, for example, any interest in those plantation and industrial forestry projects that probably bring greater benefit to wealthier capital owners, but not to poorer households. It encourages activities favoring important products for low income households but products or technologies that might first be tested and accepted by the better-off members of the local population before their less prosperous neighbors subsequently adopt them, and before the policy or practice encouraging this production achieves its full desired impact.

An Initial Set of Hypotheses

This discussion of objectives raises at least eight questions regarding the general merits of social forestry. They might be restated as hypotheses. 1) Does social forestry have a large net effect on local rural populations? The related *hypothesis 1a* is: Forest products account for large budget shares for poor rural households. (How large? Large enough for a change in product availability to alter household behavior?) *Hypothesis 1b* is: The budget share for fuelwood is greater than the share for forage or fodder, fruit or nuts. If we can not reject hypothesis 1b then fuelwood probably is the most important product of social forestry. 2) Are the distributive effects of social forestry activities (and development programs) as enlightened as we prefer to believe? *Hypothesis 2:* Social forestry activities, where otherwise justified by the efficiency criterion, yield greatest benefit to low income, generally rural households. 3) What is the income elasticity of demand for the products of social forestry? *Hypothesis 3a:* The income elasticity for fuelwood a) is large (greater than one?) and b) increases with moderate increases in income for the poorest rural households. *Hypothesis 3b:* The income elasticity for fuelwood remains large throughout all income groups in many poor rural communities. *Hypothesis 3c:* The income elasticity for forage and fodder is initially low (The very poor have no livestock.) and then increases. (Does it decrease again for even higher levels of rural income?) 4) What is the demand price elasticity for the products of social forestry? *Hypothesis 4a:* Fuelwood is price elastic a) over the ranges of price and income common to most developing countries and b) particularly for the rural poor of these countries. *Hypothesis 4b:* Forage and fodder are price elastic, yet not so elastic as fuelwood. 5) What is the substitution elasticity for the products of social forestry? *Hypothesis 5a:* There are no easy substitutes for fuelwood for the poorest populations in the most rural areas. Or wealthier and less rural populations have greater access to fuel substitutes. Alternatively, *hypothesis 5b:* Combustible agricultural residues are important substitutes for fuelwood in rural households.

A good understanding of the local household production and consumption relationships must underlie the accurate appraisal of these hypotheses. These household relationships will also show the linkages between forestry and other, primarily agricultural, activities. For example, household labor is an input to forestry and to agricultural activities. Limits on available labor may mean that labor used to collect fuelwood or other domestic consumables from the forest is unavailable for agricultural production. This would mean that the efficiency gains from some social forestry activities are best measured by their released labor, its transfer to agriculture, and the resulting expansion of agricultural production.

Savings of labor and other agricultural inputs are all the more important where agricultural products have larger budgetary impacts and are even more income and price elastic than the alternative forestry products. The common expectation is that specialized agricultural products are among the most income and price elastic, particularly for low income populations.

Household production and consumption functions may show us further distributive justifications for social forestry activities. For example, wood gathering may be mainly a women's and children's activity. If there are cultural restrictions on substituting women's and children's labor for adult male labor, then scarcer forest inventories mean that fuelwood collection requires more women's and children's labor, that less women's and children's labor is available for their specialized agricultural and household activities, and that there is a wedge between the marginal products of female and male labor in agriculture. If specialized women's agricultural activities cause a delay before the specialized men's activities can be completed, then scarcer forest resources also have the secondary effect of decreasing the marginal product of male labor. A similar pattern may be traced through the availability of specialized women's labor for fuelwood collection or food preparation, the resulting level of family nutrition, and the marginal productivity of labor in general. Decreased nutrition implies decreased productivity for both sexes in all activities.

The competition between forestry and other activities for scarce household labor inputs suggests two additional questions and their related hypotheses. 6) What are the relative efficiencies of household labor in social forestry activities and in agricultural activities? If the observed marginal products of labor are equal for all activities, and women and children are more generally involved in fuelwood collection, then (*hypothesis 6*) the marginal product of men's labor must be greater in other activities than it would be in forestry. 7) Are there cultural constraints on male-female labor substitution and are there resulting social inefficiencies? *Hypothesis 7a:* Many forestry activities are women's and children's activities. *Hypothesis 7b:* There are disproportionate social gains attached to releasing women's labor from activities like fuelwood and fodder collection and permitting their transfer to other productive activities. *Hypothesis 7c:* Cultural constraints create a wedge between men's and women's labor in forestry activities, direct men's labor away from and women's labor toward forestry activities, and result in lower marginal products of labor in forestry for women.

Our prior discussion also inquired about the introduction of new technologies into rural communities. 8) Assuming the demonstrable local efficiency of a new social forestry activity, then what is the best way to introduce it into a new community? What are the best communities in which to begin a new program? *Hypothesis 8a:* Communities with strong, receptive leaders most readily accept new ideas from outside. *Hypothesis 8b:* Economic factors outweigh cultural and political factors in explaining both initial acceptance and the continuing rate of acceptance. *Hypothesis 8c:* Households with accumulated risk capital, (perhaps larger landholding, greater wealth, or better education) more readily accept new ideas from outside.

These are all important questions. If the empirical evidence from our second and third sets of cases rejects many of these hypotheses for many general cases, this should not be viewed as a rejection of social forestry. Rather, it is an argument for careful identification of the select number of important locations and populations that do not reject the hypotheses. These are the opportunities for successful social forestry activities.

Further Implications for Publicly Supported Activities

The previous discussion is relevant to indigenous forestry activities and also to public policies and public investments in social forestry. Social forestry is not always associated with public projects or programs—but the current interest in social forestry does feature many public activities. The large public re-source management agencies and international donors that implement many forestry projects often apply two additional criteria, sustainability and public participation. These criteria are not objectives of social forestry—and they are reasonable as criteria only in certain special situations. We might reflect upon them.

Sustainability

Sustained social well-being, not sustained resource management or sustained resource outputs, is the socially responsible objective. Sustained social well-being and sustained resource management conflict, for example, in many acceptable cases of mineral development: that is, where mineral resource extraction and depletion fuel greater long-run development and overall social well-being. Forestry offers a similar example: Social well-being and sustained forest management conflict where forest depletion would permit land conversion to more productive agricultural uses. Nepal's tarai is an outstanding example. Removing the forest cover may provide only temporary, non-sustainable gains from timber harvests, but it clearly has allowed the development of Nepal's most productive agricultural region and the improved well-being of its human population.

Nevertheless, while we can have no objection to unsustainable development in one economic sector so long as it improves long-run social well-being, unsustainable forestry activities often degrade the basic land resource and make both forestry and potential agricultural development from forestland less productive. Therefore, our preferred view might be to "sustain resource options." In this case we would be less concerned with sustainable forestry *per se*, and more concerned with sustaining the land base and the diversity of the ecosystem.

Another kind of sustainability, the sustainability of public activities, should also be a source of concern for government and donor agencies involved in social forestry activities. Projects that endure only for the period of external funding can be the source of two experiences in social adjustment, one at their outset and one when the funding declines. If their dependence on the project grows large, then the local population can suffer serious hardship on the second occasion.

With this background, we can call on the efficiency and distributive criteria to identify three cases of sustainable investments in social forestry. First, where the distributive objective dominates and the final exchangeable project benefits do not exceed its project costs, then only long-term and continuing external support can sustain the project.[8] Second, and similarly, where the efficiency objective is sufficiently important but non-market values are large and market benefits still do not exceed costs, then too, only continuing external support will sustain the project. Neither of these cases is a good candidate for international loans or for eventual conversion to local public agency support. Expected project benefits cannot fully repay loans and project conversion to local support, in these cases, can occur only in conjunction with an expanding local public sector. Projects that fit these descriptions may be of great social merit but they are unsustainable without external financing. Only development agencies that are willing to provide indefinite support should accept projects that fit either of these descriptions.

In a third case, *justifiable* projects may occur where investment bottlenecks and thresholds restrict the eventual realization of benefits. In this situation, external financial assistance can provide the initial input from which sustainable benefits will follow. Education, research, extension and demonstration activities all provide new knowledge which becomes permanently endowed in the local population. Therefore, projects featuring these activities can create permanent contributions and potentially sustainable benefits. Similarly, permanent institutional changes, such as securing land tenure, can create new and sustainable management incentives. Many social forestry projects feature education, extension and demonstration components.

Others feature permanent changes in land tenure or institutional modifications within local public agencies. Two of our studies of the adoption of stoves in Nepal and the adoption of production-oriented social forestry activities in northern Pakistan reflect of the roles of education and demonstration. The tenure theme runs throughout many of our cases but cases from Nepal, China, the Philippines, and India examine this institution explicitly.

These comments on sustainability can be restated as two hypotheses. *Hypothesis 9a:* Forestry activities that are sustainable with only local resources may satisfy distributive objectives. They always satisfy strict local market efficiency criteria. This is true for i) indigenous activities and for ii) externally assisted activities that continue after the external institution and its funds withdraw from the local area. *Hypothesis 9b:* Social forestry activities collapse and an adjustment period follows which includes some local economic hardship when donor agencies withdraw their grants and loans from unsustainable activities. *Hypothesis 10:* Activities which introduce new information or improve the local skill base or alter relevant institutions in a manner that improves local resource conserving incentives have better chances to make sustainable local contributions.

Public Participation

Like sustainability, public participation is often a design requirement for community-oriented development projects. This requirement, however, is poorly defined and its usefulness in any particular social forestry project is poorly understood.

Public participation is a guide that can save sponsoring agencies from projects that are ill-designed to satisfy the target population. Its intention is to identify useful contributions to the local community. The problem is that the public participants for any given project are likely to be either the organized or the outspoken and dissatisfied among the local population. They may not even be an important segment of the target population.

There is no easy solution to this problem. The concept of public participation is easy to support but its formal organization is not so easy to define. For example, where efficiency is the key objective and markets do exist, then demand and supply analyses provide most of the necessary information. The public participates by revealing its preferences in the market. The opinions of aggressive local leaders who can venture risk capital and, thereby, invest and demonstrate the merits of new technologies, provide the remaining information.

Public participation can assume a more positive role where the distributive objective is also important and where market information is poor. But it can assume this role only if the donor's or the resource management agency's special information requirements guide the design of the formal public participation activities that seek this information. General voter measures of preference are still unsatisfactory. Project designers must listen to those among the local leaders who can provide the initial risk capital, but they must be particularly inquisitive of the budget impacts and income and price elasticities for the affected population. We previously discussed how a general understanding of these facts is essential to project design, and well-conceived public participation activities can provide them. Surveys, particularly those which intend to provide information on household income or wealth, are always difficult to design. Their results are always questionable. The importance of a narrow and specific program designed to obtain the right information from the right population is a key issue behind gathering data for several of our case studies. Our case studies will go beyond market evidence of demand and supply to search for additional demographic characteristics which help explain public acceptance of social forestry activities.

These comments on the general importance of public participation might be restated as hypotheses. *Hypothesis 11a:* Where there is an adequate market (or other means of exchange) for the products of social forestry, then formal public participation procedures only provide extraneous—and potentially confusing—information about either the efficiency or distributive objectives of public forestry investments. *Hypothesis 11b:* Formal public participation works best when designed to provide proxies for unavailable market information or to obtain explicit information on non-market values.

Secure Resource Tenure

The previous discussion prepares the way for many of our empirical case studies: for the background case studies which provide perspective on the broad and general importance of social forestry, for the production and consumption analyses which measure the importance of social forestry activities to local households, and for the broader regional assessments of demand and supply, and of the additional demographic characteristics that may affect the local acceptance of social forestry activities. Our introductory comments in this first chapter also referred to the importance of institutional change in general and changes in property rights and secure and sustainable resource tenure in particular. We will develop these institutional issues more completely in the next two sections—and in chapters 8–11.

Open access resources are resources, like many forests and grasslands, that no one owns. Therefore, no one can anticipate a secure claim on future production from the resource and no one has the incentive to invest in or to manage the resource for its future outputs. Open access resources either tend to be of very low *in situ* value or, if of higher value, they tend to be depleted rapidly as nearby citizens each rush to claim a share of the resources for themselves.

Rights of access can change, however, and when they do the new rights can act as a brake on depletion. As a region develops and its resource becomes more valuable, local custom or convention may establish more permanent rights to the future management and use of the local resources. Increasing forest values, together with better understanding and expanding concerns for forestry today, are creating the necessary sensitivities for changes in the customary rights to many local forest resources. Indeed, local citizens in many places now assert private rights for forest resources that were previously accessible to everybody. These newly established rights provide the incentive for sustainable management because those with the management rights now also obtain the future returns from their management.[9]

This sequence bears more careful consideration. During the initial change from open access, either custom or convention may justify common resource use by some well-defined local population. As development continues, the pressures of changing tastes, technology, and population levels render some customs and conventions no longer effective in maintaining these common rights of resource use. The rights establishing common management of the resource may be subject to greater pressure than the rights establishing private management because organized group dynamics may be less flexible and less responsive to external change than the preferences of single economic actors. Therefore, as development proceeds, well-managed commons may have some tendency either to break down and become the property of smaller, more responsive economic entities. For example, community forests may become the property of a few private landowners who are able to fence boundaries or otherwise establish claim to a resource that was commonly managed.[10]

This sequence from open access to well-managed commons to private property is important for social forestry because much forestland in lesser developed areas today is still either an open access or a common property resource. Forests may be the legal responsibility of the national government, but government management is often ineffective at the local level and either open access or management-in-common may be an operative fact. If the forests remain an open access resource, then the accompanying potential for depletion raises our concerns for deforestation, soil erosion, etc. If commonly managed forest resources become private, then their management incentives remain effective but the new distribution of property rights may justify our concern.[11] Commons are often the domain of poorer population groups. If the new private managers are wealthier landowners, then the poor who previously occupied the commons may be even more justifiable targets for development assistance. In this case, the same land or forest resource is a less obvious choice for social forestry projects because the wealthier new landowners are less reasonable targets for development assistance and we must find a different basis for helping the poorest households.

Finally, policy and project designers must ask whether successful development assistance will convert a poor community's common resources into likely candidates for privatization and, if so, will the conversion be long-lasting? And who are the likely benefactors? Can the policy or project design ensure that the target group of original common owners is the greatest benefactor? For example, the preferences of the local poor for projects which emphasize more immediate fodder and fuelwood returns may reflect their uncertainty regarding the permanence of any policy giving the property rights to the poor. The timber production alternative to fuelwood and fodder takes longer to develop and those of the rural poor with current access are less certain of their rights to the common forest resource in the more distant future.

Clearly, the customs and conventions explaining local property rights are especially important to long-term activities like resource-based development in general or many social forestry activities in particular. Indeed, altering or strengthening property rights can be a substantial social forestry contribution to economic development.

On the other hand, without careful empirical examination it is easy to claim too much for differences in resource tenure and property rights. Nepal's forest nationalization and more recent Community Forestry program may be an example of misplaced hypotheses. Nepal nationalized its forests in 1957 [the Private Forests (Nationalization) Act]. Most forests were previously the property of large (birta) grantholders but were managed in common by local user groups.

Arguably, nationalization is one source of Nepal's severe deforestation problem. The national government is not strong enough to securely enforce its own property rights, therefore (since 1957) local populations have treated many forests as open access resources. The result is extraction, forest depletion, and land erosion. The solution is to return some forest management responsibilities and some benefits from them to the panchayats (local communities). The government of Nepal and several donor agencies accept this solution. In 1988, five social forestry projects invested at least US $69 million in it.[12]

This argument is logically sound as it stands. It errs, however, in overlooking another, more basic, argument that the national government never had any impact beyond a few miles from the capital city of Kathmandu. Nepal has few roads and little other rapid communication even today. Therefore, nationalization may have been a meaningless act in 1957. It was an act that caused no change in most panchayats. Therefore, forests have always been and remain commonly managed resources. If this second argument is correct, then the Community Forestry/panchayat forestry solution caused no change in local action from what had occurred all along. The source of Nepal's deforestation (away from the *de facto* jurisdiction of Kathmandu) must originate elsewhere.[13]

Nepal's Community Forestry program is more complex than this brief comment suggests. We cannot deny that it is certainly more sophisticated than the many programs in many countries that plant trees without regard for property rights. Nevertheless, it should be clear that property rights are important and that rigorous empirical examination is advisable before sponsoring agencies invest scarce resources in projects based on thoughtful impressions rather than empirical facts.

This discussion raises at least four general questions about secure tenure and its management implications. These questions can be restated as hypotheses and examined empirically. 1) Are forest resources and resource services with insecure tenure causally linked with increasing relative prices for these resources and resource services? An affirmative answer means that alternative tenurial arrangements can be the solution to increasing fuelwood scarcity and forest depletion. Therefore, establishing secure resource tenure is one means of protecting the local community from high prices, resource scarcity, and economic hardship. *Hypothesis 12a:* Relative prices increase over time where land and forest tenure is insecure. *Hypothesis 12b:* Secure tenure puts a cap on rising forest product prices. *Hypothesis 13:* The security of property rights is also associated with the reliability of the policy environment establishing the rights and the landowners' expectations regarding those rights. With the rights but without confidence in their long-term possession of those rights, landowners will not make investments in long-term activities like forestry. 2) Do proposals to secure tenure satisfy efficiency criteria? *Hypothesis 14:* The net value of increased output resulting from more secure resource rights exceeds the increased costs of establishing and enforcing those rights. If this hypothesis is only reasonable in specialized cases, then what characterizes these cases? 3) Does more secure tenure create undesirable distributive losses? And are the poor the most frequent users of common property resources? *Hypothesis 15a:* The poor receive net benefits from commonly managed resources which are disproportionate to either a) their proportion of the local population, b) their contribution to common resource management or c) their share of benefits from the sum of all local resources. *Hypothesis 15b:* Securing resource tenure only transfers the resource from poor current users like squatters or shifting cultivators to wealthier holders of the new rights. Acceptance of this hypothesis raises questions about the status of poor current users once price increases justify improving the land and forest tenure—and removing resource access for the poor. *Hypothesis 15c:* The previous participants-in-common are worse off as a result.

The empirical answers to these questions are generally unknown —although there is no shortage of professional opinion. This encourages inquiry beyond our two analytical studies of commonly managed resources in Rajasthan state in India and in Nepal's mid-hills, and our reviews of emerging forest property rights in general in the Philippines or under the special conditions of policy reform in China.

Perverse Policy Spillovers

To this point we have examined the direct effects of policies and programs designed for their impacts on forestry and rural communities. We generally expect these direct impacts to be most important. Nevertheless, we might also inquire about the secondary impacts on social forestry of policies external to the social forestry sector. There is some recent argument that such "ex-sector" policies can have substantial detrimental, although unintended, impacts on forestry and rural communities.[14]

Several examples come to mind:

- High interest rates and unstable macroeconomic policies drive out long-term capital investments like forestry, leaving only unsustainable harvest activities in the remaining natural forest.
- Capital inducements (subsidies, tax concessions, specialized trade advantages) drive marginal labor from the commercial, industrial, and advanced agricultural sectors. Some of this labor migrates to rural areas, thereby increasing settlement on marginal lands and increasing pressure on the forest.
- High inflation encourages asset holding in durable resources like land; therefore inducing a bidding-up of land values, extension of the margin of developed land in order to substantiate land claims, and inefficient conversion of forestland.
- Agricultural price supports and input subsidies encourage excess conversion of forests and grassland for agriculture.
- Energy policies induce substitution away from woodfuels, but also destroy incentives to reforest.

In each of these examples, the use of trees, conversion of forests, or reduction in reforestation is an unintended policy impact. Furthermore, the larger policy benefits in each of these cases tend to accrue to better-off household, not to the rural poor.

These ex-sector policy effects do not occur everywhere. Nor are they always important where they do occur. Nevertheless, singly or as a group, some ex-sector policies do have large impacts on social forestry. Their impacts are conceivably greater than, and opposite to, the intended direct impacts of specifically-designed social forestry policies. Where this is the case, the best social forestry advice may be to change the relevant ex-sector policy.

The important questions are which ex-sector policies are important and what are the magnitudes of their efficiency and distributive effects on social forestry? These questions can be restated as multiple similar hypotheses, one for each potentially detrimental policy. *Hypothesis 16:* The particular ex-sector policy has measurable, significant, and detrimental secondary impacts on a) the forest resource and on b) rural community welfare.

More specifically, we might anticipate (*hypothesis 17*) that generally unstable macroeconomic and natural resource policy environments create investor uncertainty and discourage long-term investments like forestry. Sustainable activities, conservation, and technically efficient production are impossible without long-term investment. We might also anticipate (*hypothesis 18*) that policies that discriminate against labor and relatively labor-intensive production activities in general, or policies that discriminate in favor of capital or relatively capital-intensive production activities induce migration to the forest frontier, increase the burden on existing social forestry programs, and increase deforestation. Finally, (*hypothesis 19*) agricultural support policies are a relative disincentive for social forestry activities.

These hypotheses suggest where to look for the most detrimental ex-sector policy effects. Our chapter 2 examination of upland migration in the Philippines, our chapter 8 review of emerging forest rights, and our two case studies from China provide empirical insight.

An Appeal

Previous discussions of social forestry respond to a broad spectrum of specialized topics. Our review throughout this introductory chapter touches on some of these topics. No doubt all previous discussions have their important features. Our focus is the significant impacts of local forest-related activities on the welfare of affected rural human populations.

Our approach could be through chapters organized for each of the nineteen general hypotheses in this introductory chapter. Testing each hypothesis, however, would require substantial evidence from many and various cases. The lack of substantial evidence which encouraged us to write this book also precludes this approach. Our alternative is to rely on organized case studies which, we hope, will provide initial evidence on many of the hypotheses.

Our purpose is to encourage rigorous thinking about social forestry and the welfare of affected populations. The topic of social forestry has achieved a high level of international importance. It is entitled to the respect of scientific inquiry. Its potential human impacts demand our care. We appeal

NOTES

1. This introduction benefits from the insistence from Norton Ginsburg, David Griffin, Lawrence Hamilton, George Taylor, and Neil Byron on sharp statements, precise definitions, and clear understanding of the previous literature. The first three scholars made their comments as part of the East-West Center's generous support and hospitality in August 1987. An earlier version of this chapter appeared as "Social forestry: A working definition and 21 testable hypotheses," *Journal of Business Administration* 20(1992):430–52.

2. The development of social forestry as a topic of inquiry and the re-evaluation of plantation forestry coincided with the broader re-evaluation, in the early 1970s, of the Great Development Decade of the 1960s, and the change in the general thrust of economic development from industrial processes to "poverty focused rural development." Chenery et al. (1974) highlight this re-evaluation. N. Byron clarified this point.

3. We do not intend to disregard important applications in other countries of the region—or of the world. Space and the specific experiments of those who collaborated on this volume restrict us to some geographic selection. We anticipate that experience in other countries of the region and the world would substantiate many of our general observations.

4. One publication devoted to collecting the definitions of terms like these found 27 definitions for the one term *social forestry* (Anonymous 1985).

5. The new settlement issue is well known as a component of "transmigration" in Indonesia. It is also well known, albeit in the lowlands not the uplands, for Nepal's tarai. L. Hamilton also reminded us of the frequent distinction for social forestry between stable populations in South Asia and migrant populations in Southeast Asia.

6. Indeed, the initial conceptualizers of the dual economy understood this point well. See Boeke (1948, 1953) and Furnival (1939, 1948), or see Ginsburg (1973) for a review. Boeke's and Furnival's dual economies featured households and populations which were predominantly subsistence oriented and whose market participation was occasional. Neither their populations nor individual households relied totally on subsistence production and all participated in some market exchange. N. Ginsburg clarified this point and referred us to this literature.

7. These are empirical questions which are all too easily assumed away. For example, a large number of forest policies are justified, at least partially, on distributive grounds. Yet in the only wide-ranging assessment of the distributive impacts of forest policies known to us, Boyd and Hyde (1989) find that eight of nine U.S. forest policies redistribute from poorer in the direction of wealthier benefactors. Even the one counter-example redistributes from wealthier owners of integrated forest products firms to not-as-wealthy timber landowners. These landowners' wealth is, nevertheless, considerably above the U.S. median and further yet above any measure of poverty which normally attracts our distributive concerns. This Boyd-Hyde set of examples

reinforces our argument for thoughtful assessment before casually accepting assertions regarding distributive merit.

8. Exchangeable benefits, in this context, include benefits from goods that exchange for currency, as well as from goods that households exchange for other goods and services and goods that they hold for their own consumption.

9. The topic of property rights begs inquiry into the implicit thresholds associated with changes in the institutions of tenure. For example, how can we predict the cost levels that must be overcome in order to break down a well-managed traditional common arrangement. Or how great must private benefits be in order to sustain a new assertion of individual claims where there were no prior private claims. Or, how great must the organizational economies be in order to induce a shift from one system of rights in preference for another. No rigorous inquiries known to us examine these threshold costs or benefits.

10. Stevens (1988) tells the story of the Gurung population, high and isolated in Nepal's mid-hills who, within the last 100 years, have converted from swidden to terraced agricultural producers of staple crops. This conversion implies a shift from rotating usage of open access lands, to increased pressure and shorter rotations for the swidden crops, and finally to private claims on the land. It coincides with an increasing Gurung population, with declining swidden yields, and with exposure to lowland neighbors using different crops and production technologies. Migot-Adholla et al. (1990) tell a similar story with more formal econometrics for sub-Saharan Africa.

11. There can be social losses even if members of the same population group maintain control as individual private owners of the former commons. For example, the commons can provide a form of insurance available to participants in the commons who suffer temporary hardships. Distributing the commons among the population removes its potential to serve as a pooled insurance policy. D. Griffin provided this insight from his personal observations in Nepal.

12. This investment was equivalent to three percent of Nepal's 1986 gross domestic product. In addition, twenty more international donor activities involved related forestry, watershed, forestry training, and national park projects (Carter 1987). The World Bank alone planned another US$80 million project beginning in 1988.

13. Mahat, Griffin, and Shepherd (1986a,b) more fully explain the sequence of events having to do with Nepal's forest nationalization and its impacts.

14. deMontalembert (1991) surveys this topic. He suggests Munasinghe and Cruz (1995), organizing structures for its research.

REFERENCES

Anonymous. 1985. "Some definitions of social forestry and related concepts." Los Banos, P.I.: University of the Philippines Department of Social Forestry occasional paper no. 1.

Boeke, J. 1948. *The Interests of the Voiceless Far East*. Leiden: Universitare Pers.

Boeke, J. 1953. *Economics and Economic Policy of Dual Societies*. New York: Institute of Pacific Relations.

Boyd, R., and W. Hyde. 1989. *Forestry Sector Intervention: The Impacts of Public Regulation on Social Welfare*. Ames: Iowa State University Press.

Carter, J. 1987. "Organizations concerned with forestry in Nepal." Kathmandu: Forest Research and Information Center Occasional Paper 2/87.

Chenery, H., M. S. Ahlunwalia, C. L. G. Bell, J. H. Dulog, and R. Jolly. 1974. *Redistribution with Growth*. London: Oxford University Press for the World Bank.

deMontalembert, M. 1991. "Intersectoral policy linkages affecting the forestry sector." Washington: Paper prepared for the IUFRO-SPDC/IFPRI/USAID workshop on forestry and agroforestry policy research.

Feder, G., R. E. Just, and D. Zilberman. 1985. "Adoption of agricultural innovations in developing countries: A survey." *Economic Development and Cultural Change* 34(1): 255–98.

Furnival, J. 1939. *Netherlands India: A Study of Plural Economy*. Cambridge: Cambridge University Press.

Furnival, J. 1948. *Colonial Policy and Practice: A Comparative Study of Burma and Netherlands India*. Cambridge: Cambridge University Press.

Ginsburg, N. 1973. "From colonialism to national development: Geographical perspectives on patterns and policies." *Annals of the Association of American Geographers* 63(1): 1–21.

Mahat, T., D. Griffin, and K. Shepherd. 1986. "Human impact on some forests of the middle hills of Nepal. Part 1. Forestry in the context of the traditional resources of the state." *Mountain Research and Development* 6(3): 223–232.

Mahat, T., D. Griffin, and K. Shepherd. 1986." Human impact on some forests of the middle hills of Nepal. Part 2. Some major human impacts before 1950 on the forests of Sindhu Palchok and Kabhre Palanchok." *Mountain Research and Development* 6(4): 325–334.

Migot-Adholla, S., P. Hazel, B. Barel, and F. Place. 1990. "Land tenure reform and agricultural development in Sub-Saharan Africa." Washington, DC: World Bank, ARDD, unpublished discussion paper.

Munasinghe, M., and W. Cruz. 1995. "Economywide policies and the environment." Washington, DC: World Bank, Environment paper no. 10.

Rogers, E. M. 1983. *Diffusion of Innovations* 3d ed. New York: Free Press.

Ruttan, V. W. 1977. "The green revolution: seven generalizations." *International Development Review* 19:16–23.

Stevens, S. 1988. "Sacred and profaned Himalayas." *Natural History* 97(1): 26–35.

Part 2
Background

The two chapters in this section of our book provide an indication of the breadth of importance of forest-based activities to rural households and communities. Chapter 2 examines the attraction of undeveloped forested regions for migrating agricultural populations and the second examines the dependence of farm households on forest products. The first refers to an example from the Philippines and the second to an example from Nepal—but their issues are global. Indeed, large migration flows to newly accessible forests have been an important feature of development in Thailand's northeast, Nepal's tarai, and sparser populated areas of Indonesia, as well as the Philippines in Asia. Furthermore, population growth and agricultural conversion of the forest frontier is probably the largest source of global deforestation.

    If forest frontiers attract migration and agricultural land conversion, then what happens when the frontier declines and its forests become less accessible? We can anticipate that local households will begin to grow trees on their own lands, and for a variety of purposes. The purposes (or tree products) indicate a breadth of potential social forestry activities. It will be useful to identify the variety of forest products, and also the characteristics of households that initially manage their own lands for these products. Both the key products and the initial adopting households are important indicators—of future decisions by other local households and of potentially successful target locations and populations for forest-based development projects.

CHAPTER 2

Forest Environments as Attractants for Human Migration: The Philippines in the 1980s

Gregory S. Amacher, Ma. Concepcion J. Cruz, Wilfrido Cruz, and William F. Hyde[1]

Population pressure is generally accepted as a prime cause of ozone depletion, deforestation, greenhouse gas accumulation, pollution, and general large scale reductions in environmental quality. Migration can be a temporary outlet for population pressure, but often it only creates new population pressures and new environmental degradation in the region of in-migration.

The typical migration pattern world-wide has been from rural to urban areas, where they become a source of degradation of the urban environment. This pattern reversed itself in the 1980s, particularly in Southeast Asia, as migrants began leaving impoverished cities to settle in sparsely inhabited upland and forest frontiers. The rural poor added to the upland migrant stream as they too searched for uninhabited land and better agricultural opportunities. Government resettlement policies and policies favoring capital over labor, particularly in the industrial sector, induced disemployment and migration to the frontier where the open access nature of existing land tenure regimes provided opportunity but enhanced the likelihood of environmental degradation.

Our objective in this chapter paper is to examine Philippine migration in the 1980s, and the upland forest environment as an attractant for this migration. There are other examples—Indonesia's experience with transmigration, migration to Thailand's northeast after the army built roads for military movement and national security, or Nepal's experience with migration to the tarai after malaria was eradicated from that region—but the Philippines may be the best example. Manufacturing, employment, and per capita income actually decreased in the Philippines in the 1980s, and many poor urban and landless rural families migrated to the uplands where agricultural land conversion has destroyed 200,000 hectares of native forest annually for the last twenty years. Often this land is poorly suited for agriculture. Cultivation on slopes greater than eighteen percent, for example, increased by more than 225,000 hectares each year in the 1980s (NRAP 1991).

We will use a multinomial discrete choice model, together with census data on migration flows and socioeconomic and environmental data from the late 1980s, to assess the factors explaining the migrating population's simultaneous choices among alternative upland destinations. This approach allows us to estimate the relative importance of population factors, environmental resources, government policy, and economic opportunity at the destination as elements in the final migration decision. We will find that migrants are responsive to local income opportunity, but less responsive to population factors. The availability of upland forest and its insecure land tenure are strong and statistically significant attractants, and migration is greatest to those regions where deforestation could accumulate on the largest scale. These results argue that predictions of the effects of policies targeting employment, income, or property rights without also examining their secondary effects on the forested uplands. Similarly, predictions of the effects of trade and industrial policies that reduce the welfare of the urban poor would be incomplete without addressing the degrading impacts of these policies on the upland environment.

Migration and Deforestation in the Philippines

Poverty in the Philippines grew steadily through the 1970s and 1980s until it reached fifty and sixty percent in rural and urban areas, respectively, in 1988 (World Bank 1988). The Philippine recession of the 1980s, coupled with government incentives for capital- and energy-intensive industries, and disincentives for commercial agriculture, reinforced this disappointing trend. Displaced urban workers and landless agricultural workers migrated to the upland frontier in search of better opportunity, and they converted upland forests to subsistence agricultural production. Government forest policy—and the government's inability to enforce it effectively—reinforced the migrants' impact on the uplands. The government officially classifies lands with slopes greater than eighteen percent as forests and retains public ownership of these lands, but the government has difficulty enforcing use rights to these lands. As a result, these lands are effectively open to deforestation by improper logging and illegal homesteading.

The results have been an upland population growing at three percent annually since 1950, and settlement mostly on previously uninhabited open access forest (Cruz et al. 1988). Total forest cover has decreased by 24 percent since 1970 until 72 percent of the total upland area is now cultivated. This is more than double the area of upland cultivation in 1950 (NRAP 1991, FAO 1983).

Without secure private claims for the land, the new upland farmers tend to favor extensive cultivation practices, and short-term production maximization. The environmental results have been unsustainable agriculture on land that is poorly suited for agriculture in any event, erosion and degradation of the upland forest, and downstream losses due to heavy sediment deposition on commercial agricultural land, in water catchments intended for hydroelectric power production, and in the coastal fisheries.

The Migration Decision

The debate on world forest policy recognizes the importance of migration to deforestation, yet the empirical economic assessment of the issue is limited. Economic assessments of migration have tended to feature urban immigration, and they disregard the upland environment. Numerous geographic and demographic inquiries do examine rural immigration but this literature provides little basis for conjecture about economic policy.[2]

We will combine economic characteristics with the geographic and demographic characteristics common to previous migration assessments, but alter our focus to the migrant's choice among upland destinations, and especially to the attraction of accessible upland natural resources. An individual's decision to migrate involves a complex comparison of the known utility received at the site of current occupation with the utilities expected at various potential destinations. Once an individual chooses to emigrate, then the potential destinations form a discrete set of alternatives. Each emigrant from one origin faces the same set and the aggregate of similar choices made by individuals from one origin defines a migration stream.

Empirical Specification

Let the indirect utility for any individual with a potential destination m be given by

$$V(X_m;\Omega,\beta) + \epsilon_m, \quad \textit{for all } m \quad (2.1)$$

where X_m is a vector of variables affecting prices and income at the destination, β is a vector of estimable parameters, and ϵ_m is an error term.[3]

The probability that an individual migrates from origin province i to destination province j is

$$P_{ij} = pr\{V(X_j;\Omega,\beta) + \epsilon_j > V(X_m;\Omega,\beta) + \epsilon_m\} \quad \forall\, m \neq j. \quad (2.2)$$

We can estimate this probability with individual data (where choices are binary variables), or with sample proportions (Ben-Akiva and Lerman 1988). Sample proportions are especially well-suited for large population studies like ours where data on individuals are unavailable and where individual migration decisions can be aggregated as migration streams.

Migration streams can be defined if individuals who make similar migration choices are themselves similar. A migration stream M_{ij} is the total number of individuals in origin province i making the identical choice to migrate to destination province j. If all observations of migration are contained within one time period, then the economic, environmental, and demographic attributes that confront them only vary by the potential alternative destinations.[4]

Sample proportions define the frequency distributions of migration patterns. The sample proportion for each origin i and destination j is

$$p_{ij} = \frac{M_{ij}}{\sum_j M_{ij}} \tag{2.3}$$

The p_{ij} substitute for the P_{ij} in eq. (2.2), and the choices in eq. (2.2) become frequencies in eq. (2.3). These frequencies sum to one when they are aggregated over all migration streams from each origin province.

The error term in eq. (2.2) has an extreme value distribution (Fomby et al. 1984). Therefore, we can use a maximum likelihood procedure to determine how the probability that a migration stream occurs depends on the vector of explanatory variables in the utility function. Following Ben-Akiva and Lerman (1988), a maximum likelihood procedure follows from defining the probability of migration between origin i and destination j as:

$$p_{ij} = \frac{\exp(x_{ij}^T \beta)}{\sum_m \exp(x_{ij}^T \beta)} \quad \text{for all i,j} \tag{2.4}$$

where x_{ij} is the vector of explanatory variables from destination province j that are known to migrants from origin province i. The denominator is the vector sum over all potential destination alternatives. The probability p_{ij} is computed for all provinces, m = 1,2,...,i,j,..,N, where N is the total number of provinces. (Of course, the use of sample proportion measures of migration mean that eq. (2.4) must be corrected for heteroscedasticity prior to estimation.)

This multinomial discrete choice specification will provide consistent estimates of the effects of destination attributes on our migration streams. If we assume a Cobb-Douglas form for the representative individual's utility function, eq. (2.1), then our specification of eq. (2.4) will be logarithmic. We will follow Pudney's (1988) recommendation of linear attachments for nonprice and nonincome variables (environmental and demographic variables, in our case).

Finally, we are also interested in identifying the broad categories of explanatory variables with greatest relevance for migration. For example, we would like to know whether expected personal welfare, population and social infrastructure, or natural resource availability; or whether upland, lowland, or province-wide characteristics as a group serve as greater attractants for migration. This suggests that we should group our explanatory variables and complete our analysis with hypothesis tests that examine the importance of these groups of variables.

Data

Our fundamental source is Philippine data processed from 1990 national census observations of individuals who migrated within the 1985–1990 period. These observations were based on a stratified random sampling scheme designed to capture variation in lowland and upland socioeconomic factors. The original sample included 5,476 migration streams involving 74 provinces. A 1990 survey of the Philippine Bureau of Census provided general income, employment and demographic data. Data distinguishing upland areas within a province were taken from a University of the Philippines, Los Banos survey of 8,935 households. Annual Philippine Statistical Yearbooks provided our data on agricultural and forest lands.

We made two modifications to the data and two modifications to the analysis. First, three provinces contain no upland forests. These provinces failed to attract any immigrants and we dropped them as potential destinations. (They remain as origins for migration streams.) Second, some attribute data (income, employment, or demographic data) are missing for 19 provinces. Dropping the provinces with missing attribute data leaves 74 origin and 52 destination provinces, for a total of 3,848 observations on migration streams. This means that the model represents fewer choices, but our sample is large. Dropping a few observations will not have a serious effect on the analysis if we can assume the independence of irrelevant alternatives (IIA) (Amemiya 1986).[5] IIA is consistent with multinomial discrete choice models.

The full sample of migration streams separates into three approximately equal groups of origin-destination pairs; one for which there are no observations, the second which includes less than half the migrants from an origin province, and the third which shows that more than half of all migrants from one origin province emigrated to one destination province. The largest migration sources were the provinces of central Luzon, the most populous and industrial region of the country and the northern-most main island. Many movements were local, within the province of origin. The destinations of the largest longer distance migration streams were the 22 provinces on the less-settled and southern-most large island of Mindanao.

Table 2.1 summarizes the descriptive statistics for the destination attributes. Approximately one of every 400 individuals nationwide migrated to the uplands (line 1) to a large variety of destinations, but it is difficult to distinguish among destination attributes from the raw data. Various measures of destination attributes might be used. The measures in table 2.1 were selected to include measures of income and employment opportunity, population pressure, and resource availability. Larger total populations and larger total land area available suggest more information available to migrants and more migrant confidence about the perceived opportunity. Average income and population density are also indicative of levels of opportunity at the destination. Therefore, it is important to use each of these attributes—and not to normalize all attributes for either population or area. We will discuss our expectations for the effects of each attribute in our discussion of empirical results.

Province-wide information may be the summary information most available to long distance migrants, but the uplands tend to be their final destinations. Therefore, our data distinguish upland from province-wide attributes. Finally, several of our agriculture and forest data sets were selected for their implicit indications about secure land tenure, a critical issue for natural resource and environmental policy and management in the Philippines. Price data for selectively upland agriculture and forest resources were unavailable.

Political Unrest and Distance as Deterrents to Migration

Our analytical additions to these data are a political unrest variable and a travel cost variable. Fourteen provinces in the central Visayas and the ethnically different areas of Mindanao suffered from extreme political unrest, including guerilla activity and military responses to it. This may have made these areas less attractive. The Philippine Embassy in Washington identified these provinces and we marked them with a dummy variable.[6] A negative coefficient on this dummy would imply that political unrest restrained immigration and, thereby, deterred deforestation.

The general literature on migration anticipates that travel costs are important deterrents to migration, with larger travel costs acting as a deterrent to migrations of greater distances. Distance alone, however, does not satisfactorily incorporate the mix of time costs for overland travel and financial costs for ferry passage characteristic of migration in the Philippines. The time costs may be especially important for many poor Filipino migrants who cannot

Forest Environments as Attractants for Human Migration

Table 2.1. Descriptive Statistics for Migration Attributes

Attribute	Mean	Standard	Minimum	Maximum
Migration (% of 1980 origin population)	0.023	0.017	0.004	0.079
Average household income (pesos/yr in 1990)	31,591	8,808	16,000	55,390
Average upland household income (pesos/yr in 1990)	11,676	20,296	9,999	61,950
Unemployment rate (% in 1990)	6.16	2.75	2.0	11.9
Households in bottom 30% of income profile (1990)	12,728	3,517	7,165	23,150
Population (1980)	222,710	144,190	9,009	707,000
Upland population density (1980)	38.41	6.74	6.60	73.96
Arable land (km2)	1148.44	467.25	247	2363
Road density in arable uplands (mi/km2)	1.46	1.54	0.39	11.66
Forest land area (ha)	263,360	214,630	28,200	1,042,000
Forest land area classified as public (ha)	328.1	216.1	40.4	1041.8
Share of forest land area with >18% slope (%)	50.32	16.46	18.64	82.40

afford commercial travel. For them, the financial cost of ferry passage may be overwhelming. Moreover, the strongest felt cost of migration in the Philippines is probably its deterrent on family communication during holidays. Therefore, the threshold which makes periodic return visits unlikely is critical. This threshold may be approximated by overland travel beyond the boundary of the province of origin or by ferry travel of any distance.

We developed two alternative systems for incorporating these market and non-market features of migration costs. The first is a system of weights that assigns two points to migration streams contained within the origin province, another point to migration out of a province but within one of the twelve broad administrative/geographic regions of the Philippines, a fourth point to migration to a new administrative region, a fifth to any movement across water to a new island—or to ferry travel, and a final point to additional overland movement inland and upland once the ferry travel component of migration is complete. Some, but not all, migration out of the province includes ferry travel. Some ferry travel is followed by additional overland migration. Our second system for incorporating travel costs uses migration within a province as the standard and assigns a new dummy variable for each of the four subsequent weights in the previous system. We examined both systems, and variations upon them, in our empirical analysis.

Results

Table 2.2 displays our empirical results corrected for generalized heteroskedasticity. The first column reports the results associated with our scaled index of travel costs. The second column reports the results associated with one application of the distance variables. Except for this travel cost variable, all destination attributes are the same in both regressions, and all destination attributes are in log form. The table describes the anticipated signs and the statistical significance for each attribute, as well as the estimated coefficients. Twelve of thirteen attributes display expected signs—in both models. Ten are statistically significant in the first regression and eight are significant in the second. A Chow test rejected the hypothesis that the two regressions are significantly different.

The first group of destination attributes refers to income and employment. Average household income is a predictor of expected migrant income at the destination and, as such, is a powerful migration attractant. The uplands are the destination of most migrants, but average upland income is an insignificant predictor, probably because it represents a much smaller population and because most migrants have less well-formed perceptions of this measure than of the province-wide average household income. The income status of poorer households is also important. Migrants generally are not well-off, and it is reasonable they would be concerned with the income opportunity of the province's poorer inhabitants. The proportion of households earning less than thirty percent of average provincial household income is a measure of the opportunity of poorer households. A larger proportion earning less than thirty percent reflects poorly on migrant expectations, and deters immigration. Finally the positive sign on unemployment is surprising, but measures of unemployment are particularly unreliable in many developing countries where, and especially in rural areas where the informal sectors are particularly large. Perhaps higher unemployment rates reflect only the influx of migrants who become engaged in subsistence agricultural activities but who prefer some market employment.

Table 2.2. Multinomial Discrete Choice Estimates for Destination Attributes

Attributes		Regression	Coefficients
Average household income (pesos/yr)	(+)	3.310*** (3.708)	4.005*** (4.486)
Average upland household income (pesos/yr)	(+)	-0.011 (-0.282)	-0.0453 (-1.169)
Share of households in bottom 30% of income profile	(-)	-2.867*** (-3.280)	-2.448*** (-2.810)
Unemployment rate (%)	(-)	5.551*** (2.342)	8.003*** (3.384)
Population (1980)	(+)	1.741*** (2.932)	1.340*** (1.991)
Upland population density (1980)	(+/-)	-.0343 (-0.508)	0.505 (0.855)
Arable land (km2)	(+/-)	0.753*** (1.964)	0.384 (1.002)
Forest land area (ha)	(+)	0.577* (1.786)	0.154 (0.477)
Road density in arable uplands (mi/km2)	(-)	-0.690*** (-2.806)	-0.915*** (-3.726)
Share of forest land area classified as public (%)	(+)	0.796*** (3.719)	0.925*** (4.320)
Share of forest land area with >18% slope (%)	(+)	1.307*** (2.860)	1.414*** (3.093)
Political unrest dummy (1 for each of fourteen provinces displaying political unrest during the period 1985–90, 0 otherwise)	(-)	-0.0310 (-0.101)	
Travel cost dummy (0 for movement within the province of origin, 1 otherwise)	(-)		-54.588*** (-54.057)
Travel cost index	(-)	-3.246*** (-21.913)	
Log likelihood		-292.39	-292.39

Expected signs and t statistics in parentheses. ***, **, and * indicate significance at the 0.01, 0.05, and 0.10 level, respectively.

The second group of destination attributes refer to population. A larger provincial population probably translates into better information about the destination province held by migrants as they prepare to depart the province of their origin. (For example, Americans on the East Coast tend to know more about more populated California than they know about less populated near-by states Nevada or Arizona.) For the upland areas which tend to be the migrants' final destinations within any province, however, population density may be a more important attribute. Higher population density may indicate greater social support, an attractant, or it may indicate less available land, a detractant to immigration. Therefore, the expected sign on upland population density is uncertain—and our empirical results do not improve on our uncertainty.

The agricultural and forest attributes are indicators of resource availability. A larger arable land area may suggest greater availability, and attract migrants, or it may indicate that more land has already been settled, and detract. Our regressions indicate that the former argument must dominate. Our expectations for a larger forest land area are unambiguously positive. Larger forest area implies more land available for settlement, and greater opportunity for deforestation and conversion to new small-scale upland agriculture. The regression coefficients support these expectations.

The remaining agricultural and forest attributes all reflect on the security of existing claims to the land. Less secure existing claims are migration attractants because they indicate greater opportunity for immigrants to extract some claim for themselves. Of course, the resulting new claims cannot be very secure, and short-term management and environmental destruction will probably accompany new settlement. Lower road densities in arable uplands suggest either that the existing farms are larger and less-populated with tenants, or that the uplands contain a larger unroaded and undeveloped interior. In either case, areas that are less roaded would be more susceptible to settlement by new immigrants. The government is largely unsuccessful at restricting settlement on public forestland. Therefore, the public share of the forest land base is also a migration attractant. The share of forestlands with steeper slopes is officially all public. It is an attractant and an indicator of likely off-site environmental damage made certain by the insecurity of a settler's tenure on forestlands that are nominally public. This insecurity encourages migrants to make temporary and unofficial claims but it discourages long-term investments and conservation management practices by the new upland settlers.

The distance and political unrest dummy variables performed as anticipated. Distance and political unrest both detract from migration streams, although the detraction due to political unrest is statistically insignificant and small in magnitude. Perhaps political unrest sends a mixed signal, indicating distress which is a migration deterrent, but also indicating less security of existing land tenure and, in a few cases, attracting potential settlers who will compete for the tenurial rights. On the other hand, political unrest and local concerns for land tenure are mutually reinforcing, and the insignificance of our political unrest dummy variable may be due to the interdependence of these two issues.

The first column of results in table 2.2 reports the coefficient for our travel cost index. It is negative and highly significant—as expected. The regression with four distance dummies did not perform well, but it did indicate a critical threshold at the borders of the province of origin. Perhaps this is because the migration streams identified by the second through fourth dummies (migration out of the province to various more distant destinations) were too small for reliable statistical analysis. When we re-estimated (effectively lumping these dummies) such that our only distinction is between migration within, or beyond, the province of origin, then this second regression performs well. The second column of table 2.2 reports the results with this dummy variable. (The larger coefficient on this distance dummy is due to the dummy's smaller value: 1 or 0 for the dummy in column two versus weights between 2 and 7 for the distance variable in the first column of results.) These results show the importance of the border of the origin province as a migration threshold and, thereby, lend support to the argument that the strongest felt cost of migration is its deterrent to family communication during holidays.

Hypothesis Tests

A series of hypothesis tests can broadly identify which sets of attributes are most important to large-scale migration patterns. Travel cost is clearly a key variable, but our greater interest is in the relative importance of the various destination attributes. We can examine the relative contributions of sets of attributes by dropping a set, re-estimating the model, and conducting a likelihood ratio test with the initial and re-estimated models. Our likelihood ratio tests would reject the null hypothesis that attributes in the set are unimportant if the value of the test statistic is large.

Table 2.3 reports the likelihood ratio test statistics when each of several attribute sets is removed from the first regression in table 2.2. We cannot reject the null hypothesis for the sets of income, population, and general resource availability attributes. The test statistic supports the importance of the two poverty attributes (households in the bottom thirty percent of income, and unemployment rate) and of the attributes that reflect upland characteristics alone (upland income, upland population density, roaded arable uplands, government forestland). The test statistic for the poverty attributes supports our expectation that the status of poorer households reflects closely on migrants' expectations for their own opportunities. The test statistic on upland attributes argues that, while general province-wide income, population, and general resource availability attributes are unimportant, more selective upland attributes are important for the choice of a migration destination.

Table 2.3. Hypothesis Tests (Values of the Likelihood Ratio Test Statistic)

Attribute Set	Destination Attributes Dropped from Model	Test Statistic
Income	(avg. income, avg. upland income)	0.036
Population	(population, upland pop. density)	0.036
Poverty	(bottom 30% of income profile, unemployment)	20.05*
General resource availability	(arable land, total forest land)	1.88
Upland resource availability	(upland pop. density, roads in arable uplands, public share of forestland)	67.84*

* indicates significance at the 0.05 level or better

Indeed, the insignificant coefficient on upland income in table 2.2, together with the first (income) and fifth (upland) hypothesis tests, encourage a variant on the latter with upland income removed. (The last row of table 2.3 reports this new Chi-square test statistic.) Altogether, these observations seem to argue that upland resource availability is a more important migration attractant than expected income. This would be a most reasonable finding for immigrants whose subsistence agricultural opportunities are more important than their participation in the cash economy. This last hypothesis test is critical. It confirms our initial expectations that it is the uplands that are attracting migrants in the Philippines, and that regions which display lower population density and larger areas of insecurely tenured lands are especially attractive. Settlement of these areas implies deforestation, and the difficulty the settlers have in establishing their own long-term rights to these lands implies a preference for short-term management practices that would create potentially important erosion and off-site environmental destruction.

Conclusions

Evidence from Philippine migration in the 1980s confirms our expectation that attributes associated with expected opportunities at the destination sites determine migration choices among alternative destinations—and it highlights the importance of accessible natural resources. Upland income and employment opportunities may be attractive, but the availability of undeveloped land like forests and (developed or undeveloped) land with insecure rights to the existing tenure is particularly important.

This is a crucial insight for policy analysis, and especially for social forestry. Land at the frontier tends to be more fragile and more susceptible to environmental destruction. This is certainly the case for the steep upland frontiers of the Philippines. The combination of fragile lands and insecure tenure raise grave concern for the possibilities of excessive deforestation and for downstream damage to reservoir catchments, prime agricultural land, and the in-shore fisheries. Policies that correct the tenure problem at the forest frontier will have important positive environmental impacts, especially where the immigration flow is substantial. Policies designed to encourage capital, subsidize energy, or support commercial agriculture or international trade often have unintended negative impacts on employment. Therefore, they unintentionally encourage migration and expand environmental damage at the frontier. The proponents of these policies seldom consider their indirect impacts at the frontier or on deforestation, but our results suggest that those impacts can be great where the migration streams are large.

NOTES

1. Forestry Department, Virginia Tech; Global Environmental Facility, World Bank; Environment Division, World Bank Institute; and Centre for International Forestry Research; respectively. An earlier version of this paper appeared as "Environmental motivations for migration" *Land Economics* 74(1)(1998): 92–101.

2. Cruz et al. (1988) compute changes in upland forest populations in the Philippines and informally tie population growth to the decrease in forest area. Todaro (1976) reviews earlier economic migration models. Cruz and Francisco (1993) review migration and the environment.

3. Moving costs incurred by the migrant are implicitly present in the indirect utility function because moving costs affect income.

4. This focus on groups of migrating individuals, rather than the total population, and on the attributes of destination locations, is appropriate for examining choices among potential migration destinations—but not for assessing the question of whether or not to emigrate. It is especially appropriate for our interest in the different environmental characteristics associated with upland destinations.

An assessment of the migrant's prior decision of whether to migrate would require a prior examination of the attributes of migrating individuals in comparison with the attributes of non-migrating individuals in the origin population. This would require data on non-migrant populations which are unavailable to us. If these data were available, the destination choice could be nested in the decision to migrate.

5. We tested for IIA by removing three of the largest provinces and computing the appropriate test statistic (Hausman and McFadden 1984). We could not reject the hypothesis of independent destination choices.

6. L. Q. Del Rosario, personal communication, October 11, 1995.

REFERENCES

Amemiya, T. 1986. *Advanced Econometrics*. Cambridge: Harvard University Press.

Ben-Akiva, M., and S. Lerman. 1988. *Discrete Choice Analysis: Theory and Application to Travel Demand*. Cambridge: MIT Press.

Cruz, M. C., I. Zosa-Feranil, and C. L. Goce. 1988. "Population pressure and migration: Implications for upland development in the Philippines." *Journal of Philippine Development* XV(1).

Cruz, W. C., and H. Francisco. 1993. "Poverty, population pressure, and deforestation in the Philippines." Washington: Environment Department, World Bank. Draft manuscript.

Fomby, T., C. Hill, and S. Johnson. 1984. *Advance Econometric Methods*. New York: Springer-Verlag.

Food and Agriculture Organization. 1983. *Tropical forest resources assessment project: Forest resources of tropical Asia*. Rome: FAO.

Hausman J., and D. McFadden. 1984. "Specification tests for the multinomial logit model." *Econometrica*. 52(5): 1219–1240.

Natural Resource Accounting Project (NRAP). 1991. "Final report of phase I activities." Manila: Department of Environment and Natural Resources and US Agency for International Development.

Pudney, S. 1988. *Modeling Individual Choice*. Cambridge: Basil Blackwell.

Todaro, M. 1976. "Internal migration in developing countries: A review of theory, evidence, methodology, and research priorities." Geneva: ILO.

World Bank. 1988. *Philippines: The Challenge of Poverty*. Washington.

CHAPTER 3

Some General Features of Household Forestry in Nepal

K. H. Gautam, B. R. Joshee, R. L. Shrestha, V. K. Silwal, M. P. Suvedi, and L. P. Uprety with William F. Hyde[1]

This chapter extracts from a series of local studies to reflect on the general occurrence of trees on private lands and the demand for tree products by local households in Nepal. Its observations are true indicators of the breadth of forestry's importance for rural households because they reflect the households' willingness to use their own scarce labor and agricultural lands to grow forest products.

The authors of this chapter each conducted household surveys for purposes of their own and from panchayats (villages) of their personal knowledge. Their surveys are not a representative sample of all Nepal, but they do provide consistent and detailed impressions across a variety of communities in different economic and physical conditions. The combined observations from these surveys are suggestive of the relationships explaining household production and consumption of fuelwood, fodder, fruits and nuts, and construction wood. In addition, three of the surveys allow us to reflect on the characteristics explaining the acceptance of new forestry activities. Altogether, this chapter raises questions that anticipate our second and third sets of case studies on household production and consumption, on adoption, and on regional supply and demand. Finally, comments on public versus private land as sources of fuelwood and fodder appear occasionally throughout the chapter. These comments anticipate the discussion on property rights, open access, and common property featured in the fourth and fifth sets of case studies later in this volume.

Background

Nepal is one of the world's six poorest countries. Fifty-five percent of its population earns less than the subsistence standard of sixty dollars per capita. Nepal is an agricultural country, but 46 percent of its population is landless or near landless and the average household owns only 0.4 hectares.

Nepal contains three distinct geographic regions each running the 800 km length of the country: the mountains above 4000 meters, the mid-hills ranging from 1000 to 4000 meters in elevation, and the tarai, the local name for the Gangetic plain. The cool and moist hills are the traditional population center. They support a population density of 10.53 persons per hectare of arable land. [This compares, for example, with only 8.23 persons per hectare in Bangladesh (Bhandari 1985).] Nepal's population concentration has shifted toward the warmer and drier tarai, however, since the eradication of malaria in that region in the 1950s permitted immigration and conversion of the tarai forest to productive agriculture. The tarai now supports approximately half of Nepal's total population—but more than forty percent of the tarai's forest cover has been removed for either agricultural conversion or domestic wood consumption. The removal rate for forest cover is 2.1 percent annually for the country as a whole.

Clearly, Nepal is an appropriate focus for our concerns with economic development and rural poverty. Indeed, foreign assistance to Nepal approximating one-fourth of gross domestic product, and including several social and community forestry projects, acts on exactly these concerns.

Household Production and Consumption of Tree Resources

Forests satisfy several primary consumer demands in Nepal. We might look to the local forests for indications of local production patterns and indications of the scarce resources that households are willing to forego in order to satisfy their consumer demands. Many forests, however, are under government management or the *de facto* management of local village elders. Other forests are open access resources. In either case, exclusion from the forest is difficult and the forest is largely available to the entire local population. Therefore, the condition of the natural forest is not reflective of the long-term production incentives facing local households. Measures of privately grown trees will be more reliable indicators of these incentives. In one summary measure of the importance of private forests, Joshee (1986) estimated that they provide for 24 percent of household energy consumption.

Even the most densely inhabited private lands in the poorest regions support some trees, occasionally on private woodlots but more often as small stands along terrace banks and walls, property boundaries, irrigation channels, or in the immediate vicinity of farm households.[2] Individually standing trees can provide a large share of the total consumption of forest products. In Bangladesh, for example, Byron (1984) estimated that ninety percent of all wood products originate from the scattered trees on smallholder farm land. In Malawi, Hyde and Seve (1993) projected that smallholder tree plantings would provide for all domestic fuelwood consumption within ten years.

Shrestha (1986), Silwal (1986), and Gautam (1986) (SSG) surveyed households in one tarai and five hill panchayats to assess the private ownership of trees and the importance of forest products for each household. Observations from four of their five hill panchayats may reflect an upward bias from the general case because these four are within the regions of donor-assisted forestry development projects. Nevertheless, an upward bias is still informative because it reflects landowner receptivity to forest development projects and reinforces the observation of landowner demand for trees and tree products.

Table 3.1 reports the SSG findings. SSG divided trees into six product classes: fuelwood, fodder, fuelwood and fodder combined, timber, fruit, and bamboo. Many trees produce multiple products: for example, an annual fodder crop until the trees mature or until their leaf production begins to decline, at which time they become sources of fuelwood, or even construction timber.

Each SSG product class corresponds to a basic household necessity. Wood is the primary source of heating and cooking fuel for most households. Heating demand is greater in the cooler hills than in the tarai and in the winter months than in the summer. All households cook at least one meal daily but cooking demand is greater among wealthier households which cook two meals daily. Fodder sustains a domestic livestock population that is among the highest in the world: 15.6 million buffalo, cattle, goats, sheep and pigs; including 4.4 bovines per household in the hills and 6.2 per household in the tarai (World Bank 1974). Open grazing opportunities are available during the monsoon season between June and September but fodder and crop residues must supplement grazing during the remainder of the year, and particularly in the winter months of October to March. Fodder and crop residues are the only food sources for stallfed animals. These are the strongest draft animals and they also furnish larger volumes of meat and milk per animal than rangefed animals. Poles for supports in domestic housing construction are the primary timber product, and bamboo is a secondary source of poles for construction. (Only Silwal included bamboo clumps in his survey.) Most fruit production is used to satisfy direct domestic consumption.

These general observations anticipate Shrestha's findings. Shrestha separated her observations by three household land ownership classes. She found that larger landowners uniformly possess more trees for each use, although all classes of landowners own some trees, including even the smallest landowners in the warmer tarai where fuelwood for heating is less important and the land's agricultural productivity is greater. Fruit-bearing trees are important in both the hills and the tarai. Shrestha found that other uses (fuelwood, fodder, and timber) are more important in the hills than in the tarai. The number of trees per household initially rises approximately proportionately with area of land owned, and then rises more than proportionately for most kinds of trees and most landowners. Hill landowners own 40–400 trees of all

Table 3.1. Number of Trees Owned per Household by Purpose

Panchayat location and identification (researcher)	Tarai #1 (Shrestha)	Hill #1 (Shrestha)	Hill #2 (Shrestha)	Hill #3 (Silwal)	Hill #4[a] (Silwal)	Hill #5 (Gautam)
Average farm size in hectares[b]	1.1–3.7–12.1	0.3–0.7–2.1	0.3–0.7–2.1	0.5	0.5	unknown
Tree purpose						
fuelwood	0.0–0.3–2.4	0.3–6.4–40.0	1.9–40.6–216.3	65.5[c]	17.9[c]	39.0
fodder	0.5–1.1–1.0	7.9–15.7–51.8	4.8–7.6–34.1	1.5	2.0	33.1
fuelwood and fodder	0.4–0.0–0.0	22.7–24.1–45.9	1.7–5.6–126.1	–	–	–
fruit	4.5–12.9–47.8	8.8–9.3–45.8	4.1–7.4–11.8	14.7	5.3	–
timber	0.7–1.8–12.6	2.8–6.3–56.3	1.9–2.6–50.5	[c]	[c]	37.1
bamboo clumps	–	–	–	3.6	2.9	–

[a]Not within the region of a forestry development project
[b]Shrestha separated her survey respondents into small, medium, and large landowners
[c]Fuelwood and timber combined

kinds per hectare. In sum, it is difficult to say which tree product is more important overall but Shrestha's evidence suggests that trees are clearly an important land use in both hill and tarai communities.

Silwal's and Gautam's observations generally support Shrestha's observations, although without reference to landowner size. Silwal's second panchayat (hill panchayat #4 in table 3.1) is the lone hill panchayat in our sample that is not also within the region of a forestry development project. Since the survey sampling procedure was identical, the difference between Silwal's hill panchayats #3 and #4 probably reflects the positive impact of the development project on tree ownership in panchayat #3. Nevertheless, even households in that panchayat that is not part of a development project report almost thirty trees of all purposes per hectare.

Shrestha's survey respondents relied less on common and public lands for fuelwood and fodder at the time of her survey than five years prior to it. None of her respondents relied solely on the common lands for fodder, and private production of both fuelwood and fodder increased over the five years prior to the survey. Market purchases of fuelwood also increased. Tarai respondents tended to purchase their wood products while hill respondents tended both to purchase some and to grow some.

Trees that replace agricultural crops on the household's own land, and household purchases of tree products with scarce cash reserves, are indicators of the importance of tree products in the life of rural households. The level of effort that households extend to obtain fuelwood and fodder and the expectations displayed by local experiences in planting trees are further indications of their importance.

Shrestha's hill survey respondents spent three hours per bhari (headload) collecting fuelwood and consumed 10–15 bharis (25–35 kg/bhari) per household per month. They spent another hour per headload of fodder collected and they collected one-to-three headloads daily. Together, fuelwood and fodder collection occupied the time of approximately one-half person per household per day in the hills.[3,4] The demand on household time was greater in Shrestha's tarai panchayat where more than one person per household per day may be involved in fuelwood and fodder collection.

Fuelwood collection increases with household wealth, probably because wealthier households cook a second meal (increased demand), have more land and trees (increased supply), and tend to have larger families. Larger families suggest both more consumption and more able bodies to assist with fuelwood collection. Fodder collection increases with numbers of livestock owned—which also increases with the size of the land holding and with family size. Therefore, fodder collection also correlates positively with wealth.

Shrestha observed little seasonal variation in time spent collecting fodder. Near-by private lands tend to be the source for summer collection, while the more distant public or common lands may be the source for winter collection. Summer is a peak season for agricultural activities. Winter is a drier season. Therefore, more labor is available for collection from greater distances and walking those distances is easier in the winter.

Table 3.2 reports local experience planting trees and local desires to plant more trees. Its cells duplicate table 3.1 except for the addition of a new column reporting Uprety's (1986) observations from a sixth hill panchayat (also within the region of a forestry development project). Once more, neither landowner perceptions nor landowner actions leave any doubt about the importance of trees. Even in the tarai and in a panchayat unassociated with a forestry development project, and even for the smallest landowners whose agricultural production must be most limited yet most critical to personal livelihood, 32 percent of households have planted trees and 79 percent desire to plant more. The numbers are higher for larger landowners and in the hills.[5]

Adoption of New Technologies

If the merit of planting is clear to these Nepali households, then which of them are most likely to plant additional trees. Or, similarly, which household characteristics best predict the likelihood of early household adoption of new technologies and the acceptance of new social forestry projects?

The general economic argument is that those households with a plentiful supply of risk capital are the first to try new technologies. Their more risk averse neighbors observe, and then follow the successes of the first group of risk takers. This general argument is difficult to test: Wealthier and better-educated households may possess risk capital and may adopt first—or they may only be the natural and easy targets for information relayed by the extension agents who have the responsibility to transfer information about new social forestry opportunities.

In order to address these issues, Joshee examined initial acceptance of a new technology, improved stoves, Shrestha investigated the comparative household awareness of information from alternative sources (*i.e.*, extension agents or radio news) and Suvedi compared characteristics of mobile and stationary households. Mobility suggests a willingness to take risks. Therefore, it may be evidence of risk capital. The characteristics of mobile populations also may be the characteristics of innovators and early adopters of new technologies.

Joshee surveyed fifty households in each of two hill panchayats twenty kilometers north of Kathmandu. We might anticipate a general incentive for these households to adopt more efficient stoves. Ninety-eight percent of Nepali households heat and cook with fuelwood and crop residues. Moreover, fuel imports account for 66 percent of Nepal's foreign currency expenditure.

Table 3.2. Perceived Shortages and Planting Intentions (numerical measures are either trees planted or percent of landowners)

Panchayat location and identification (researcher)	Tarai #1 (Shrestha)	Hill #1 (Shrestha)	Hill #2 (Shrestha)	Hill #3 (Silwal)	Hill #4[a] (Silwal)	Hill #5 (Gautam)	Hill #6 (Uprety)
Average farm size in hectares	1.1–3.7–12.1	0.3–0.7–2.1	0.3–0.7–2.1	0.5	0.5	unknown	unknown
Tree purpose				trees planted per household in 5 years		percent of landowners	95 percent of all landowners have planted trees for some purpose within 20 years
fuelwood	percent who have planted			–	–		
fodder	32*–56–80	83*–61–80	94*–71–83	5.0	12.4	83	
fuelwood and fodder	percent who want to plant more			46.5	2.0	87	
fruit	79–94–100	82–78–79	79–64–75	–	–	–	
timber				11.5	5.1	100	
bamboo clumps	–	–	–	2.7	1.8	–	

[a] Not within the region of a forestry development project

* All those who planted trees, planted on their own lands. Those entries (from Shrestha's observations) without an * may have planted on either public or communal lands as well as their own lands.

Forty-six percent of the population suffers from respiratory ailments—often traced to open stoves and to smoke in the home. All of this anticipates a strong interest in increasing fuel efficiency. Nevertheless, improved stoves have not been adopted widely.

Joshee grouped his observations by households that a) adopt and continue to use improved stoves, b) adopt, then abandon improved stoves, and c) continue using traditional stoves. Tables 3.3 and 3.4 report the mean values of his observations. They leave us uncertain about the middle group who adopted, then abandoned the improved stove technology. It is clear, however, that size of landholding, livestock ownership, family size, off-farm employment, literacy, and current school attendance are all greater for households that have adopted and continue to use the improved technology. All of these characteristics are indicators of greater wealth and they probably indicate greater accumulations of risk capital. Table 3.4 supports this contention. Households adopting the improved technology have larger incomes and, particularly, larger off-farm incomes. They also spend more for all purposes, including fuel.[6] The fact that those households that continue to use traditional stoves have more unaccounted net income per capita may only reflect hoarding to protect against the unidentified risks that these less wealthy households can least afford.

Joshee's observations are consistent with the hypotheses that wealthier households have more risk capital and those with more risk capital are more likely to adopt new technologies. Bhandari (1985), however, offers the counter-argument that extension agents and social workers are more likely to seek out higher caste large landowners. Therefore, it is not surprising that extension activities transfer information about new technologies to better-off households first.

Shrestha (1986) tested Bhandari's argument by investigating landowner awareness of the Community Forestry Development Project (CFDP). The two sources of information about CFDP are the Forestry Department and the media—radios and newspapers. Shrestha argues that radio news is equally available to all households. Therefore, the observation that households seem equally aware of CFDP through each information source indicates that the Forestry Department is not just talking to wealthier households and that there is no extension agent bias in her sample population. Wealthier landowners must be more responsive to external information from whatever source.

If Shrestha's argument is convincing, then we might continue the search for indicator characteristics of initial innovators. We might search where the local power elite are not well-established; therefore where they have little impact on adoption. We might also search in mobile populations because mobile populations have already shown themselves to be alert to new

Table 3.3. Household Characteristics and Stove Preference[a]

Stove type	Land holdings	Livestock units[b]	Family Size	Agricultural employment	Off-farm employment	Literacy	Children attending school
Improved	0.9 ha	139	11.0	2.5	2.0	3.4	3.7
Abandon improved	0.3 ha	5	6.8	3.0	1.4	2.2	2.1
Traditional	0.4 ha	26	8.0	2.6	1.2	1.7	3.0

[a]Source: Joshee (1986) survey of 50 households in two hill panchayats 20 km north of Kathmandu.
[b]Raw total of cattle, buffalo, goats, sheep, pigs, chicken.

Table 3.4. Household Income and Expenditures (in rupees) Compared with Stove Preference[a]

Stove type	Income				Expenditures				Net income per capita
	Agric.	Off-farm	Fuelwood	Total	Agric.	Off-farm	Fuelwood	Total	
Improved	4,756	24,333	—	29,089	6,411	15,561	838	22,810	578
Abandon improved	3,611	15,100	—	18,711	3,563	6,920	289	10,872	1,167
Traditional	2,833	9,793	55	10,843	2,031	5,323	367	8,742	2,650

[a]Source: Joshee (1986) survey of 50 households in two hill panchayats 20 km north of Kathmandu.

opportunities. The condition of mobile populations without well-established community leaders describes many of the newer tarai panchayats.

Suvedi (1986) surveyed the poorest twenty percent of households in comparable rural tarai and hill panchayats. He found that poor hill households are more aware of the various health and agriculture programs, but the only adopters of new technologies are among his tarai households. Some poor tarai households planted improved varieties of grain and some relied on health clinics rather than traditional medical remedies. Shrestha's survey supports the contention that mobile populations are younger, more literate, have smaller dependant households and have higher off-farm incomes. There is greater ethnic diversity in Shrestha's poor tarai population—and it is more likely to adopt new technologies.

Conclusions

The observations in this chapter are only an introduction to many of the subsequent chapters. Nevertheless, the sum of the evidence in this chapter is convincing. Trees produce a variety of important products for rural Nepali households, regardless of whether the households are very poor or not-so-poor, and whether they are in the warmer and drier tarai or in the cooler and moister hills.

Trees are important producers of a range of goods for rural domestic consumption: fuelwood for heat and cooking, fodder for livestock, timber for construction, and fruit for direct human consumption. While few agricultural households possess their own woodlots, very many of even the smallest landowners plant and grow substantial numbers of trees wherever they can. This is true both in the hills where heating is an important fuelwood use and the tarai where warmth is less an issue and the land has a greater productive potential in agriculture.

Nearly all landowners surveyed have planted and wish to plant more trees on their own lands. This observation is consistent with the large share of household labor occupied in fuelwood and fodder collection [approximately fifteen percent of total household labor (Joshee 1986, Shrestha 1986)]. The considerable landowner interest in trees increases, perhaps even more than proportionately, with growth in the size of the landholding or implicit household wealth. Nevertheless, the demand for tree products is probably income inelastic. Tree products are probably basic economic necessities and their relative scarcity is increasing. (Evidence on the relative price trends that could confirm this conclusion was not collected in these surveys.)

The evidence of planting, as well as the stated preferences of landowners to plant additional trees, both suggest that larger landowners and households in the cooler hills of Nepal will be most responsive to future social forestry activities. The evidence of improved stove adoption confirms the expectation that larger households may be the first to adopt. It also adds household wealth and education as indicators of adaptive behavior. Tarai households have been more mobile than hill households and they are also more ready respondents to new public programs. Therefore, mobility may be another indication of adaptive households in general, and tarai households they may be more responsive than hill households to new social forestry opportunities. In general, however, the evidence in this chapter supports the contention that social forestry has some role for most households and in all regions of Nepal.

This chapter concentrates on the forestry decisions of private landowners. Nevertheless, it contains some information about tree planting opportunities on lands held in common. Some survey respondents alluded to planting the commons. Others refused to use their own labor to plant the commons because they mistrust the local leaders who may exercise greater claims on these public resources. On the other hand, the landless must depend on the commons and all households may rely on commons for fuelwood and especially fodder in the drier winter months. Several later chapters will return to these themes of property rights and commonly-owned forest resources.

NOTES

1. District Forest Controller, Dalakha; Economist, APROSC, Kathmandu; Senior Economist, APROSC; Assistant Research Officer, National Commission on Population, Kathmandu; Lecturer, Tribhuvan University, Rampur; and Assistant Professor, Tribhuvan University, Kathmandu; and Senior Associate, CIFOR; respectively, at the time this research was completed.

2. In his survey of landowners in one panchayat, Gautam finds that only the very largest private landowners grow trees in identifiable forest stands. All other landowners grow trees singly or in small groups in locations that permit minimal interference with agricultural production.

3. Joshee (1986) observed the higher rate of sixteen percent of all household labor occupied in fuelwood collection in two towns twenty kilometers north of Kathmandu.

4. Paudyal (1986) found that households that are actively involved in providing fuelwood for urban markets in Bhaktapur (a city adjacent to Kathmandu) spend another one-half day delivering it. His evidence suggests that the poorest large households and lower caste families participate in market fuelwood production. Their incentive for spending the additional time delivering firewood is the higher fuelwood price in this urban market and, therefore, the greater return on their labor opportunity. (These very poor households also have less land and fewer agricultural opportunities for their labor.)

5. It is an interesting observation that Shrestha's smallest landowners never participated in planting on communal or public lands, yet larger landowners do

sometimes participate in the planting of common and public lands. Do small landowners believe they will receive little benefit relative to their effort? Or do they possess little residual effort which they can risk contributing to the common good? Or both? Silwal's farmers would probably answer affirmatively to the first: they anticipate little personal benefit. Silwal finds evidence of farmer mistrust for panchayat leaders, and this mistrust causes independent farmers to participate less in communal activities.

6. As income increases, fuelwood expenditures decrease from approximately five percent to approximately three percent of total income. This observation is consistent with the reasonable expectation that fuel is a necessity. Its income elasticity is much less than one. Joshee's observations of household labor occupied in fuelwood collection suggest a similar conclusion regarding its income elasticity. Labor occupied in fuelwood collection increases with family size and with income, but at a declining rate.

REFERENCES

Bhandari, B. 1985. "Landownership and social inequality in the rural tarai area of Nepal." Madison: University of Wisconsin unpublished PhD dissertation.

Byron, R. N. 1984. "People's forestry: a novel perspective of forestry in Bangladesh." ADAB News (Association of Development Agencies in Bangladesh) 11(2): 28–35.

Gautam, K. H. 1986. "Private planting: forestry practices outside the forest by rural people." Kathmandu: Winrock International forestry research paper series no. 1.

Hyde, W. F., and J. Seve. 1993. "The economic role of wood products in tropical deforestation: The severe experience of Malawi." *Forest Ecology and Management* 57(2): 283–300.

Joshee, B. R. 1986. "Improved stoves in minimization of fuelwood consumption in Nepal." Kathmandu: Winrock International forestry research paper series no. 7.

Paudyal, K. 1986. "Noncommercial cooking energy in urban areas of Nepal." Kathmandu: Winrock International forestry research paper series no. 4.

Shrestha, R. L. 1986. *Socioeconomic factors leading to deforestation in Nepal.* Kathmandu: Winrock International research and planning paper series no. 2.

Silwal, V. K. 1986. "Attitude, awareness, and level of people's participation in the community forestry development project, Nepal." Kathmandu: Winrock International forestry research paper series no. 3.

Suvedi, M. P. 1986. "Forest of the poor: a comparative study of rural poverty in two villages of Nepal." Kathmandu: Winrock International rural poverty research paper series no. 3.

Uprety, L. P. 1986. "Fodder situation: an ecological-anthropological study of Machhegaon, Nepal." Kathmandu: Winrock International forestry research paper series no. 5.

World Bank. 1974. "Agriculture sector survey of Nepal." Washington: World Bank report no. 519a-NEP.

Part 3
Household Production and Consumption, and the Adoption of New Technologies

The household is the basic decision making unit in rural villages, and fuelwood is probably the most common forest product. Fuelwood is critical for heating and cooking because oil, gas and kerosene generally do not penetrate the rural markets. Fuelwood collection is often a woman's activity, and this raises questions about gender and equity. As fuelwood becomes scarcer, households may substitute lower quality combustible agricultural residues for energy, and this suggests a long-run decline in soil productivity and sustainable agriculture. Many forest development policies and projects respond to these issues with community forestry activities, or by distributing new seedlings or improved stoves or other energy conserving technologies.

The economic literature on these issues, however, is undeveloped and the empirical evidence is sparse. The chapters in this section begin the discourse with fuelwood evidence from Nepal and Pakistan. The first chapter contrasts household behavior in two districts of Nepal's mid-hills. It provides estimates of fuelwood production and consumption elasticities, income elasticities, and substitution opportunities with combustible agricultural residues and improved stoves. It examines the productive impacts of women's and men's labor, and it examines the changes in fuelwood production that occur with increasing resource scarcity on the district's common forest lands. Its evidence cautions against easy conclusions about womens' roles and encourages confidence in the ability of local households to adjustment in the presence of real economic scarcity due to deforestation.

This chapter, and the second chapter with evidence from Pakistan's Northwest Frontier Province, go further to examine the characteristics of those households which are most likely to respond to new social forestry opportunities. These characteristics should indicate the best target populations for successful social forestry programs. The second chapter also provides evidence that other aspects of the technology transfer, in particular the character of the project personnel advising on the new technologies, are as important as the characteristics of adopting households themselves.

CHAPTER 4

Household Fuel Production and Consumption, Substitution, and Innovation in Two Districts of Nepal

Bharat R. Joshee, Gregory S. Amacher, and William F. Hyde[1]

Much of the interest in forestry and rural development has focused on fuelwood, its production and consumption, and their effects on the poorest households and on the environment. The usual arguments are that primarily women and children collect fuelwood for household consumption, combustible crop residues (e.g., rice straw) substitute for fuelwood in the poorest households, and fuelwood consumption is a normal good (*i.e.*, has an income elasticity between zero and one) while residues are inferior goods (*i.e.*, have negative income elasticities) for poor households. Therefore, forestry activities that increase the availability of fuelwood and development projects that promote improved stoves both release women's labor from fuel collection and permit its use in other productive activities and also improve the agricultural environment by permitting crop residues to be better used for enriching depleted soil. These arguments are most encouraging for those of us concerned with deforestation and deforestation and with equitable development opportunities for women and the very poor. Unfortunately, they are largely unsubstantiated beyond casual observation.

This chapter is one attempt to assess these arguments rigorously and quantitatively. It relies on a survey of 99 households from two districts in the east-central hills of Nepal, Sindhupalchok and Kavreplanchok.[2] The labor opportunity costs of collecting fuelwood from more distant *common* forestlands are greater in Sindhupalchok. We will find that women's role in fuelwood collection varies importantly between Sindhupalchok and Kavreplanchok, and the male role also becomes important in the district where fuel is scarcer. Domestic fuelwood production on *private* agricultural lands in this district is one piece of evidence that Nepali households respond rationally to increased resource scarcity.

Our analysis begins with a utility maximizing household. A representative household uses various labor and capital inputs in household production and consumption decisions involving fuelwood, food crops, and combustible residues. The household is responsive to fuel prices, labor costs, an alternate fuel consumption technology, and farm and non-farm income. Empirical specifications of our model for the two different districts can be used to examine a) the importance of community forests for fuelwood collection, b) the relative importance of alternate labor sources (male or female, adult or child, family or hired), c) jointness in fuel and food crop production, d) substitution in consumption between fuelwood, residues, and improved stoves (a technological substitute for fuel), e) the importance of household activity (farm or non-farm) or demographic characteristics like ethnic group or education in explaining consumption patterns, and f) the magnitudes of fuelwood and residue price and income elasticities.

The chapter continues with an examination of the adoption of improved stoves. Improved stoves are enclosed replacements for the traditional open tripods familiar in poor households throughout the world. Enclosed stoves consume less fuel and reduce the pressure to deforest. They also funnel their smoke discharge, thereby reducing human exposure to a prime cause of the upper respiratory diseases that affect 46 percent of Nepal's population, and they indirectly improve nutrition for the entire household by reducing the time women spend collecting fuel, thereby increasing the time they have for food preparation.

Our analysis of this technology transfer problem also begins with a utility maximizing household that must decide, first, whether to adopt an improved stove and, then, how efficiently to use it. Our empirical results from the same Kavreplanchok and Sindhupalchok survey suggest the economic and demographic characteristics of those households that lead their communities in adopting the improved technology. This analysis differs from other studies of adoption in that a) it examines a consumption technology under uncertainty, b) the technology can vary in efficiency, and c) technology adoption has an unknown wealth effect.

Household Production and Consumption

Consider a representative household that maximizes a twice differentiable quasi-concave utility function which depends on the consumption of fuelwood, food crops, crop residues, and other goods. The household allocates family and hired labor to fuelwood collection and agricultural production, and it produces crop residues and some fuelwood in the same fields with its food crops. The household can purchase or sell fuelwood and food crops in complete and competitive markets, but its purchases are constrained by the sum of household profits from the sale of fuelwood and food crops plus exogenous household income from business and service activities.

More formally, the representative household maximizes utility by choosing the optimal combination of various labor L_i and capital K inputs, and the optimal bundle of consumer goods Y_j

$$V(p,I;C) = \max U(Y_j;C) \qquad j=f,c,r,o$$
$$\text{s.t.} \quad \theta(L_i,K;Q_j) = 0 \qquad i=h,d$$
$$L_\Sigma = L_h + L_d$$
$$L_d + R^\circ = t \qquad (4.1)$$
$$\Sigma_j(Q_j-Y_j) + M = wL_h = 0$$
$$L_i, K, Y_j, Q_j \geq 0$$

where V(.) is indirect household utility. p is a vector of market prices, I is household income (a function of parameters in the budget constraint), and C is a vector of exogenous characteristics that influence household behavior. U(.) is household utility and the subscripts f, c, r, and o refer to fuelwood, food crops, residues, and other household consumption goods, respectively. θ is a general aggregate production technology that relates labor and capital inputs to household production Q_j. Total labor L_Σ is the sum of hired labor L_h and family labor L_d. (Hired labor includes hired male L_{hM} and hired female L_{hF} laborers. Family labor includes adult male L_{dM} and female L_{dF} household members; and male and female children, L_{dm} and L_{df}, or L_{dc} together.) Family labor plus household leisure time R equals total time t available to the family. Leisure is fixed at R° in our model. The household's marketed surplus is $(Y_j - Q_j)$, M is exogenous income, and wL_h is the wage bill for hired labor.

Fuelwood production is really a complex set of alternative technologies—and household choices between these technologies distinguishes between the two Nepali districts. We can refine the general production technology in (4.1) for 2 cases: 1) Households may collect fuelwood from forested *commons*. In this case, it is generally argued that the labor of children and female adult family members is the sole input. The production function θ may be rewritten as

$$\theta_1[\phi_1(L_d;Q_f), \psi(L_i,K;Q_c,Q_r)] = 0 \qquad (4.2a)$$

where $\phi_1(.)$ is the fuelwood collection technology and $\psi(.)$ is a joint production technology for crops and residues. 2) Households also may collect fuelwood from trees and woody plants growing on the household's *private* agricultural lands. In this case, all categories of labor may participate in fuelwood production and collection. Adult males plant and care for young seedlings, occasionally using some forms of agricultural capital (e.g., spades, draft animals for cultivation), and they may later collect fuelwood on their way to and from the fields. In this case, the production function may be rewritten as

$$\theta_2[\phi_2(L_i,K;Q_f),\psi(L_i,K;Q_c,Q_r)] = 0 \qquad (4.2b)$$

Of course, L_i and K are generalized in θ_2. ϕ_2 and ψ do not use the same L_i and K. The first case is more descriptive of Kavreplanchok where the commons are larger and more accessible, and the second case is more descriptive of Sindhupalchok. Food crops and crop residues are jointly produced in the fields of both districts, with inputs of capital and all sources of labor.[3]

Household profits from the sale of agricultural products and fuelwood can be written as $\gamma[p_j, \theta(L_i,K;Q_j)] = \pi(w,r)$ where r is the rental rate of capital. The household budget constraint in eq. (4.1) can be rewritten (from this profit function and the constraint on leisure time) as

$$\Sigma_j p_j Y_j = w(t-R^n) + \pi(w,r) + M \qquad (4.3)$$

Complete markets exist for labor, fuelwood, and all agricultural crops, and we assume that households are indifferent between family labor and hired labor. This means that household production may affect household demand through the affect of production on household profits, but household demand has no affect on household input allocation and household production. This form of "separability" simplifies our empirical analysis by allowing us to exclude consumption variables from the assessment of household production (Singh et al. 1986).

With a separable model structure, consumer demands Y_j^D can be derived from the constrained indirect utility function defined in eq. (4.1):

$$\begin{aligned} Y_j^D = -V_{yj}/V_{Pj} &= (U_{Yj} - \lambda p_j)/(-\lambda Y_j) \\ &= Y_j^D(p_j, p_{k \neq j}, M, \pi(.), w; C) \end{aligned} \qquad (4.4)$$

where λ is the Lagrangian multiplier associated with the budget constraint and the subscripts Y_j and p_j on V indicate partial derivatives. Consumer demand decreases in response to own price increases, and increases in response to price increases (or quantity decreases) for substitute goods. Therefore, we anticipate that residue demand increases in response to fuelwood price increases, and that fuelwood and residue demands decrease for owners of improved stoves.

Three characteristics of fuelwood and residues require special attention in empirical assessments of the production and demand functions, eqs. (4.3) and (4.4). First, some households purchase all of their own fuelwood. They collect none. This implies that the opportunity wage bill from other allocations of these households' own labor may exceed their implicit return from fuelwood collection. Conversely, for those households that collect but do not purchase fuelwood, the opportunity wage bill from other labor allocations may be less than the market return from fuelwood collection.

Second, agricultural residues do not exchange in local markets. This does not violate the separability condition because residues are a by-product of crop production, and households allocate their labor and capital inputs to crop production in a manner that is consistent with local markets for these inputs. Jointness in food crop and residue production, yet a local preference for using family labor in residue collection, may suggest scale economies in residue production for those larger households also employing hired labor in crop production. Furthermore, if residues are an inferior good and fuelwood and residues are substitutes in consumption, then both scale economies in residue production and the inferior nature of residue consumption support the equity argument for development programs with a fuelwood component.

Finally, various household characteristics may affect demand: Larger families (who cook more) and farmers (who heat food for their large animals) may consume more fuel. The household's ethnic background, education, or income source may reflect specialized tastes or access to information that can alter an individual household's demand for fuel.

Empirical Analysis

A 1986 survey of 99 households from two geographically distinct districts in hills of east-central Nepal provides the data to investigate household behavior. The survey includes observations on fuelwood, crop, and residue production, purchase, and consumption over a one-month period; various categories of labor participating in fuelwood and residue collection; household income; and other household characteristics (stove quality, animal ownership, income source, education, ethnic group).

Several characteristics of these data are important for our choice of estimation technique. First, each district represents one competitive market. Price variation across households within each district is small, but we expect significant price variation between districts. The distinction between districts will be important for understanding our fuelwood production and residue demand regressions. Second, the absence of input cost data forces us to examine production functions rather than the more usual cost functions and associated input demand functions. Third, not all households both collect and purchase fuelwood. Therefore, our data include some zero-valued observations, a condition that suggests tobit models for our regressions. Finally, the large observed differences in household size and income level are potential sources of heteroskedasticity in the error terms of our production and demand functions. Madalla (1983 ch. 3) recommends linear or quadratic approximations of the unknown functional forms to correct the tobit estimates for heteroskedasticity.[4]

Fuelwood Production

Table 4.1 shows our fuelwood production regressions for the two districts. These regressions are empirical counterparts to the fuelwood component ϕ of eqs. (4.2). Each regression explains monthly fuelwood collection in kilograms as a function of the number of laborers of each category who participate in collection. Two regressions include a measure of household inputs of agricultural capital.

Table 4.1. Tobit Estimates of Fuelwood Production[a]

Independent variables	Kavreplanchok		Sindhupalchok	
	commons	private	commons	private
constant	55.37**	64.97*	-6.14	-24.99*
	(6.25)	(4.77)	(-0.01)	(-2.36)
hired male labor, L_{hm}		-7.20		-2.52
		(-0.74)		(-0.20)
hired female labor, L_{hf}		-11.67**		14.06*
		(-1.74)		(2.48)
male child labor, L_{dm}[b]	60.74	9.50		
	(0.49)	(1.22)		
female child labor, L_{df}[b]	-74.30	-1.53		
	(-0.56)	(-0.20)		
child labor, L_{dc}[b] (combined male and female)			-17.83	65.54*
			(-0.01)	(4.90)
adult male labor, L_{dM}		11.21*		37.22*
		(2.99)		(8.08)
adult female labor, L_{dF}	9.95*	6.50*	25.78	9.69*
	(4.39)	(2.09)	(0.07)	(3.12)
non-lactating bovines (proxy for agricultural capital, K)		0.16		6.97*
		(0.05)		(3.19)
significance level of a Lagrange multiplier test for heteroskedasticity	0	0	0	0
log of the likelihood function	234	246	324	302
degrees of freedom	41	42	44	39

[a] Results are corrected for heteroscedasticity in household size and income. Numbers in parentheses are asymptotic t ratios.

[b] Male and female child labor are separated in Kavreplanchok where there are many observations, but combined in Sindhupalchok where there are few observations.

*Significant at the 0.01 level, **significant at the 0.05 level.

The first (commons) regression for each district explains our anticipated fuelwood collection function from a community forest.[5] The second (private) regression for each district explains fuelwood production on and around the household's private fields. The general expectation is that women and children are important participants in fuelwood collection in both cases. We expect that hired laborers collect fuelwood only in the latter case. Some agricultural capital used in food crop production also may be an important input to long-term fuelwood production in and around the household's fields in the second case. Agricultural capital would not be important in fuelwood collection from community forests.

Our production regressions support these expectations. The area around our survey households in Kavreplanchok is more abundantly forested and we anticipate that households in Kavreplanchok are reliant on community forests for their fuelwood. Kavreplanchok's households have not yet begun growing many trees on private agricultural land. Therefore, it is not surprising that the commons production regression performs well for Kavreplanchok, while the additional independent variables in the private production regression make no apparent contribution to our understanding of fuelwood collection in that district. Households in Sindhupalchok are more reliant on private production and the second regression performs better for them. Indeed, the commons regression is virtually meaningless for Sindhupalchok. Our proxy for agricultural capital in the private regressions is non-lactating bovines, many of which are draft animals. This proxy is an important predictor of the level of household fuelwood production in Sindhupalchok. Tobit likelihood tests are satisfying for both the first (commons) regression for Kavreplanchok and the second (private) regression for Sindhupalchok. Appendix 4A reviews the evidence confirming input-output separability and joint crop-residue production in Kavreplanchok.

Fuelwood price and fuelwood collection time evidence from the two districts further support our explanation of community forest collection for Kavreplanchok and largely private fuelwood production for Sindhupalchok. That is, if there is one market price in each district, then a longer collection time for the marginal load of fuelwood would suggest a greater supply price in the local market and a smaller level of fuelwood consumption per household in that district. At some point, the greater market price would become sufficient to justify households forgoing some (non-fuel) agricultural opportunities and producing fuelwood on their own private agricultural land. Therefore, if Kavreplanchok relies on community forests, but the community forest inventory is less plentiful in Sindhupalchok (where households must be more reliant on private production), then we would anticipate longer collection times, a greater price, and less consumption in Sindhupalchok.

Indeed, this is exactly what we do observe. The mean collection time for Sindhupalchok households is 4.7 hours per headload, while the mean collection time for Kavreplanchok is 3.75 hours per headload. (The difference is significant at the 0.01 level.) The mean market price reported by Sindhupalchok households is 13.3 rupees per headload, while the mean price in Kavreplanchok is 11.7 rupees per headload. (The price observations in Kavreplanchok are too few to obtain a statistically meaningful measure of the difference.) If the two districts are otherwise alike, then farm households in Sindhupalchok have greater incentives to grow their own fuelwood. Finally, mean household consumption in Sindhupalchok is 116 kg per month, while consumption in Kavreplanchok is 134 kg per month. (The difference is significant at the 0.05 level.)

Perhaps the most important policy observation to take from this production evidence is that women and children are not the sole collectors of fuelwood. Children play a mixed role, perhaps because they require a level of supervision that detracts from total household effort spent on fuelwood collection. (Apparently, fuelwood collection and childcare are joint products of women's labor.) Women play an undeniable role in fuelwood collection, particularly when the fuelwood originates from somewhere away from the household's own fields.[6]

Nevertheless, our regressions reveal that the marginal fuelwood contribution of adult male family members is greater than the marginal contribution of adult female family members when the fuelwood originates from the household's own fields.[7] This finding may be unexpected, but it is reasonable. Men spend more time in the fields. Therefore, they are likely to carry to the household a larger share of the products, whether agricultural crops or fuelwood, from their own fields.

The contributions of hired labor are more difficult to explain. Perhaps hired male laborers travel from their own homes to their employer's fields and return, but they seldom travel to their employer's homes. Therefore, they seldom transport fuelwood for their employers. Hired female laborers, however, both work in their employers' fields and assist in their employers' homes. Therefore, hired females do have occasion to carry fuelwood from their employer's fields to their employer's homes and this explains the significant positive coefficient on L_{hF} in the private regression for Sindhupalchok.

If we can generalize from the experiences of Kavreplanchok and Sindhupalchok, then development activities that increase community forest production may still leave fuelwood collection in the hands of female labor. Activities that focus on areas with higher fuelwood prices and larger shares of private fuelwood production may generate greater household benefits—but the benefits accrue to both male and female laborers. Our Kavreplanchok and Sindhupalchok evidence has a second broad policy implication. It suggests that some level of deforestation is eventually self-correcting. Or, as deforestation and fuelwood scarcity reach some level, fuelwood consumers begin planting trees on their own private lands.

Fuelwood Expenditure

Table 4.2 shows the fuelwood consumption regressions for the two districts. These regressions are empirical counterparts to eq. (4.4). We have no reason to expect that the consumption patterns in the two districts are different. (A likelihood ratio test does not reject the hypothesis that the two functions are comparable.) Therefore, we can lump our survey data and estimate household fuelwood demand for the two districts combined. The final column of table 4.2 is our primary evidence. Table 4.2 also shows the results of independent estimates for each district, and for separated high and low income groups in the combined districts. The high and low income regressions will permit a more thorough examination of income elasticity.

Each regression estimates monthly fuelwood consumption in kilograms a function of fuelwood price and cost variables, household use of improved stoves, two household income sources, and two demographic variables. Tobit estimation is unnecessary for the high income regression because the observations are complete for high income households. All evidence of significant heteroskedasticity was corrected in each equation and the final equation test statistics are encouraging in all five regressions.

Only 25 of 99 households in the combined districts reported observations of fuelwood prices. This limited number of observations and the minimum variation in these observations anticipated with only two competitive markets (one for each district) both suggest that fuelwood price would be an unreliable predictor in our survey. On the other hand, 83 of 99 households collect but do not purchase fuelwood. For these 83, the implicit gains from other labor allocations are less than the return from collecting fuelwood, and collection time (the time in hours that it takes to collect the marginal headload) is a good measure of the household opportunity cost of fuelwood. We can use it as a proxy for price.

Our fuelwood demand regressions support the use of the collection time variable.[8] The collection time coefficient is significant in four of five regressions and it has the anticipated negative sign in all five. The collection time elasticity estimates are always small and lower income households are even less responsive to variations in collection time. These observations further support the contention that fuelwood is a basic necessity.

The coefficient on improved stoves is a measure of the technical substitution of stoves for fuelwood. (Kerosene and other non-biomass fuels are largely unavailable in the rural hills.) Four of the five regressions confirm our

Table 4.2. Tobit estimates of fuelwood expenditure[a]

Independent variables	Kavre-planchok	Sindhu-palchok	Combined districts		
			Low income	High income[b]	Combined
constant	87.49* (3.53)	78.19* (3.21)	82.89* (3.64)	-4.96 (-0.18)	48.85 (1.20)
fuelwood collection time (proxy for price, p_t)	-6.78** (-1.90) [-0.17]	-7.12* (-3.60) [-0.27]	-6.26** (-1.70) [-0.28]	-0.84 (-0.19) [-0.84]	-5.03* (-3.70) [-0.157]
improved stove	-10.75 (-0.59)	-12.08* (-2.23)	-14.24** (-1.93)	22.01 (0.70)	-33.31* (-5.38)
household profits from agricultural production, π	0.0017 (0.79) [0.12]	-0.0019 (-0.88) [-0.20]	-0.0060* (-3.14) [-0.31]	0.00048 (0.37) [0.00048]	0.0042** (1.69) [0.348]

exogenous income, M	0.0015 (1.28) [0.07]	-0.0060* (-2.78) [-0.26]	-0.0066* (-3.17) [-0.20]	0.0019 (1.23) [0.0019]	-0.0032* (-5.51) [-0.135]
family size	3.65 (0.72)	13.72* (4.81)	15.68* (3.65)	14.14* (5.37)	20.56* (2.04)
Newar		-17.18 (-0.22)	-21.68 (-0.97)		4.36 (0.50)
significance level of a Lagrange multiplier test for heteroskedasticity	0	0	0	b	0
log of the likelihood function	264	259	244	269	618
degrees of freedom	40	39	41	44	88

[a] Regressions corrected for heteroskedasticity. Parentheses indicate asymptotic t ratios. Brackets indicate elasticities taken at the mean.
[b] OLS estimates in log-log form because there are no censored observations on the dependant variable. adj R²=.54, F=26.76.
* Significant at the 0.01 level, ** significant at the 0.05 level

expectation that the substitution effect dominates and improved stoves reduce fuelwood demand. The income effect may be stronger for higher income households. The stove coefficient suggests that the improved technology reduces household fuelwood consumption by 10 kg or eight percent in Kavreplanchok, by 12 kg or ten percent in Sindhupalchok, or by 33 kg or 29 percent in the combined population. These are satisfying observations for policy makers who encourage the adoption of improved stoves as a means of reducing fuelwood demand and, therefore, forest depletion.[9]

The coefficients on agricultural profits and exogenous income are generally either negative or small and insignificant, but they eventually become positive for higher income households. The estimated income elasticities support a contention that fuelwood is an inferior good, particularly for lower income households. The larger elasticities for agricultural than for non-agricultural income encourage the proposition that additional agricultural opportunity increases fuelwood consumption. Perhaps this only means that households with more land also have more trees.

The general inferior nature of fuelwood consumption may have favorable long-run implications for the local environment. As the economy develops and household incomes rise, lower income households may eventually behave more like our higher income households. In this event, Nepali households will consume less fuelwood and draw less on the indigenous forests. Therefore, local forestry development projects may be most important for providing a near-term fuelwood supply and for protecting the forest environment until the configuration of demands on the forest change.

Finally, we considered various demographic variables. Only one is ever statistically significant. Larger families consume more fuelwood. They heat larger homes and cook more meals per household. Newars are an ethnic group that consumes additional fuel while brewing their traditional alcoholic beverage. Their additional consumption, however, does not have a significant impact on demand. Neither do various indicators of animal demand or household education.

Residue Demand

Table 4.3 shows the residue consumption regressions for the two districts independently, and for high and low income households for the combined districts. We will discuss a residue production effect on consumption that distinguishes between Kavreplanchok and Sindhupalchok. A likelihood ratio test supports this distinction by rejecting the hypothesis that the residue demand functions are similar for the two districts. The high and low income residue demand functions are presented only because they are instructive about income effects.

Table 4.3. Tobit estimates of residue demand[a]

Independent variables	Kavre-planchok	Sindhu-palchok	Combined districts[b]	
			Low income	High income
constant	11.63 (0.35)	47.59 (0.67)	-13.11 (-0.83)	38.86* (3.01)
fuelwood collection time (proxy for price, p_f)	5.80* (2.18) [0.62]	-10.19* (-4.39) [-0.52]	5.44* (2.24)	2.10 (0.98)
improved stove	-2.18 (-0.07)	-69.77* (-4.76)	-6.44 (-0.52)	-30.62* (-2.05)
household profits from agricultural production, π	-0.0020 (-0.63) [-0.59]	-0.00036 (-0.76) [-0.09]	0.0024** (1.67) [0.36]	-0.00039 (-0.64) [-0.00039]
exogenous income, M	-0.0033* (-2.62) [-0.63]	-0.0016 (-0.26) [-0.17]	0.00061 (0.39) [0.05]	-0.0012* (-1.69) [-0.0012]
family size	5.48 (0.95)	15.54* (2.45)	1.52 (0.74)	1.66 (1.32)
significance level of Lagrange multiplier test for heteroskedasticity	0	0	0	0
log of the likelihood function	221	274	217	233
degrees of freedom	40	41	43	44

[a] Regressions corrected for heteroskedasticity. Parentheses indicate asymptotic t ratios. Brackets indicate elasticities taken at the mean.
[b] A likelihood test rejects the hypothesis of similarity between the Kavreplanchok and Sindhupalchok demand functions. Evidence from the combined demand functions is instructive for the π and M coefficients only.
* Significant at the 0.01 level; ** significant at the 0.05 level

These regressions are empirical counterparts to eq. (4.4). Each regression explains monthly consumption of combustible crop residues in kilograms as a function of the fuelwood opportunity cost, household use of improved stoves, two household income sources, and a demographic variable; but not as a function of own price. Residues are seldom traded and local markets for residues do not exist in either Kavreplanchok or Sindhupalchok. A plausible explanation is that residues, as joint outputs of agricultural production which are only collected by family members on their way home from the fields, may have a very small labor opportunity cost. They are consumed only if their substitutes are expensive or unavailable. Finally, all evidence of significant heteroskedasticity was corrected and the final equation test statistics are encouraging in all four regressions.

Consider the independent variables in turn. The fuelwood opportunity cost, or collection time, is our proxy for the price of substitutes.[10] Table 4.3 shows that the increased fuelwood collection time from the community forests of Kavreplanchok induces a higher level of residue substitution for fuelwood. This substitution is more pronounced for poorer households and it is insignificant for higher income households. These income class observations are consistent with the expectation that residues are an inferior good.

Fuelwood collection time has an unexpected negative affect on residue demand in Sindhupalchok. Its explanation lies in the production differences between the two districts. The many Sindhupalchok households growing their own fuelwood would already gather what residues they can as they collect fuelwood from their private agricultural lands. It is difficult for them to substitute any more residues for fuelwood.

It would be reasonable that those remaining Sindhupalchok households that do collect fuelwood from community forests have little agricultural land of their own. These households would grow little of either food crops or fuelwood. Their relatively greater food demand per hectare probably causes their production of food crops to crowd out the private opportunity for fuelwood production. Furthermore, their lower levels of crop production also imply a low level of residue production. These smaller farmers collect few residues and they must travel whatever distance it takes to collect their fuelwood. This explanation would create the negative coefficient for the impact of collection time on residue consumption in Sindhupalchok.[11]

The stove coefficient has the negative sign anticipated for technological substitutes. Since unused residues are returned to the soil, then improved stoves are an important soil sustaining and environment conserving technology. The stove coefficient is larger for higher income households and for Sindhupalchok. The larger coefficient for higher income households is yet further support for the contention that residues are an inferior good. The larger coefficient for Sindhupalchok reflects our previous observation of greater fuelwood scarcity in that district.

The general expectation is that only poor households are substantial consumers of residues. The income coefficients in table 4.3 support this expectation. The coefficients on agricultural profits and non-farm income are positive only for low income households. They are small and negative in all other cases. These negative coefficients, and their increasing negativity with increasing income, are the convincing evidence that residues are inferior goods. Observations of the non-farm income coefficient are particularly informative as they are uncontaminated by any agricultural production effect. The non-farm coefficients are even more negative than the coefficient for agricultural profits.

Finally, larger families, as expected, consume more agricultural residues. No other demographic variable is important for predicting residue demand. Newar households do not use residues in alcohol preparation.

The Adoption of Improved Stoves

Our fuelwood and residue demand regressions both argue that improved stoves are important technological substitutes for fuel consumption, yet we know that not all households adopt improved stoves. We also know that the introduction of improved stoves is a favorite activity of conservation-oriented rural development projects. Nevertheless, there seems to be little understanding of where improved stoves are likely to be adopted, or the projects are likely to be successful. We might improve on this condition by examining the characteristics of households that adopt improved stoves beginning with a model of household utility maximization. We can examine our model empirically with evidence from our Kavreplanchok-Sindhupalchok survey from Nepal.

The Household Model

For household decision makers confronting a new technology, the uncertain gains from adoption contrast with the certain financial cost of the new technology, an improved stove in our case. The uncertain gains vary with the efficiency of household use of the stove. Thus, both the decision to purchase an improved stove and the choice regarding its level of use depend on household attitudes toward risk and household expectations of the gains from its adoption.

Consider a representative household's decision to purchase an improved stove. The household maximizes a utility function with arguments for the consumption of fuelwood and of other goods. Both the fuelwood production technology and the household budget constrain consumption. A portion of fuelwood collected may be sold in an existing market. The household's problem is to select its labor and capital inputs and its level of fuelwood consumption, as well as to make a decision regarding the purchase of a stove. Once the household decides to purchase a stove, then it must also choose the best level of stove use. The complete household problem is

$$V(p,I;C) = \max E[U(Y_f, Y_o; C)]$$
$$\text{s.t.} \quad \theta(L,K;Q_f) = 0 \qquad (4.5)$$
$$p_o Y_o + p_f[Y_f - \alpha(e,S)] + S + w(R-t) - \pi(w,r) - M = 0$$
$$Y_i, Q_f, L, K, R, M \geq 0$$

where maximization occurs over fuelwood consumed Y_f and fuelwood produced Q_f, other consumption goods Y_o, and labor L and capital K inputs. V(.) is indirect expected utility and U(.) is household utility, as before. p_f, p_o, I, and C remain fuelwood and other consumption good prices, income, and exogenous household characteristics, respectively. We assume that U(.) is separable in fuelwood and other goods; thus $U(Y_f, Y_o; C) = U^1(Y_f) + U^2(Y_o; C)$. θ(.) is the production technology, S is investment in an improved stove, R is leisure time, t is total time, w is the wage, and r is the rental rate of capital. π(.) is the profit function from sales of fuelwood and other household production, as before.

Household income has a deterministic component M (exogenous income), and an unknown component [-$p_f\alpha$(e,S)] representing the decrease in fuelwood expenditure when S is invested in an improved stove. Thus, the certain outlay of S introduces a random wealth term into the household budget. The function α(.) represents two characteristics of the stove adoption decision. First, in order to reduce fuel consumption, the household must invest in (adopt) an improved stove. Second, increasing investments in S imply increasing levels of stove efficiency.[12] The more the household invests in the stove, the greater the deterministic component of household income. The form of α(.) includes most classes of uncertainty and the economic intuition is the same in all classes. The multiplicative form αSe is illustrative.[13] The parameter e is a random variable such that $E(e)=\beta$ and $VAR(e)=z$, where $\alpha(e,S) \geq 0$ for all S, and $\int \alpha(e,S)dS = 1$ where S is in [0,∞). We assume that $\partial \alpha(.)/\partial S \geq 0$ and $\partial^2 \alpha(.)/\partial S^2 < 0$.

We can optimize with respect to the investment in stove efficiency S by removing fuelwood consumed Y_f from eq. (4.7). Solving the budget constraint for Y_f and substituting into eq. (4.7) allows us to restate Y_f in terms of random wealth W(α,S), where

$$W(\alpha,S) = \left(\frac{1}{p_f}\right)[w(t-R) + \pi(.) + M - p_o Y_o - S] + \alpha(e,S)$$

Ignoring the non-negativity conditions for now, the new household problem is

$$V(p,I;C) = \max E[U(W(\alpha,S),Y_o;C) - \lambda_1\theta(L,K;Q_f)] \quad (4.6)$$
$$= \max E[U^1(W(\alpha,S);C) + U^2(Y_o;C) - \lambda_1(.)]$$

where maximization now occurs over S, Y_o, L, and K. λ_1 is the Lagrangian multiplier associated with the household's production function. Note that $U[W(\alpha,S),Y_o;C]$ is quasi-linear in $W(.)$ and Y_o. The second line in eq. (4.8) represents the static household decision to adopt an improved stove technology. In reality, the household's decision is intertemporal, since current investment in an improved stove affects future fuel consumption. Therefore, a more appropriate intuitive explanation is that our model represents the discounted value of the infinite stream of reductions in fuelwood consumption due to a one time investment of S.

Household consumption and production decisions are separable in line two of eq. (4.8). This means that uncertainty regarding improved stove adoption has no impact on the household's allocations of labor and capital inputs, and that we can focus on the fuelwood consumption and improved stove decisions.

The first order conditions for the optimal investment in an improved stove and the optimal level of consumption of other goods are

$$V_s = EU_w^1(.)\left[\frac{\partial\alpha}{\partial S}(e,S) - \frac{1}{p_f}\right] = 0 \quad (4.7)$$

$$V_o = E[U_w^1(.)\left(-\frac{p_o}{p_f}\right) + U_o^2(.)] = 0 \quad (4.8)$$

where the subscript "0" refers to partial differentiation with respect to Y_o. We assume that the second order conditions hold such that $V_{ww} \leq 0$, $V_{oo} \leq 0$, and $V_{ww}V_{oo} - (V_{ow}V_{wo}) > 0$.

Conditions (4.9) and (4.10) can be used to examine the two aspects of the household's improved stove decision: adoption and the level of use. In order to determine a condition for adoption (S>0), we can first identify the corner solution (S=0) and then observe when it is not satisfied. After setting S=0 in conditions (4.9) and (4.10), condition (4.10) becomes irrelevant and condition (4.9) can be expressed as

$$V_s = U_w^1[W(\alpha,0),Y_o;C]E[\frac{\partial \alpha}{\partial S}(e,0)-\frac{1}{p_f}] \leq 0 \qquad (4.9)$$

If S=0, the sign of V_s in condition (4.11) must be strictly negative. Therefore, the household adopts an improved stove if the expected marginal decrease in fuelwood consumption is greater than the inverse fuelwood price. The probability of adoption (S>0) approaches one as the fuelwood price approaches infinity. In the particular case where $\alpha(.)=\alpha Se$, the household will adopt as long as $\beta \geq 1/p_f\alpha$.

If the household adopts, then S>0 and conditions (4.9) and (4.10) simultaneously determine the household's investment in stove efficiency. We can improve our intuition about stove efficiency by holding consumption of other goods Y_o constant. Rewriting condition (4.9),

$$E[\frac{\partial \alpha}{\partial S}(e,s)-\frac{1}{p_f}] = -COV[U_w^1(.),\frac{\partial \alpha}{\partial S}(e,S)-\frac{1}{p_f}]/EU_w^1(.) \qquad (4.10)$$

From condition (4.11) and the assumption that $U_w(.)\geq 0$, the LHS of condition (4.12) must be less than or equal to zero when $S \geq 0$. Thus, condition (4.12) implies that the household chooses a stove such that the expected return from owning it (the reduced household value of fuelwood consumed) equals the covariance between the marginal value of wealth changes (U_w) and the marginal income change [$\partial\alpha(e,S)/\partial S$]. Or, at the chosen level of S, the expected gain from having the stove (LHS) equals the covariance between marginal utility and marginal income. We expect that this covariance is negative because, as the marginal return from investing in a stove increases, the term $\partial\alpha(e,S)/\partial S$ also increases; while the marginal utility of wealth must decline if utility is concave.

Once more, where $\alpha(.)=\alpha Se$, the household optimally invests S* such that condition (4.12) becomes

$$\alpha\beta = -COV[U_w^1(.),\frac{\partial \alpha(e,S)}{\partial S}]/[EU_w^1(.)]+\frac{1}{p_f} \qquad (4.11)$$

When stove investment is strictly positive (that is, the household has already chosen to invest), conditions (4.9) and (4.10) can be examined simultaneously to determine the affects of exogenous parameters on the level of stove investment.[14] Conceivably, these affects depend on the household's attitudes toward risk: is it increasingly risk averse, decreasingly risk averse, or risk neutral.

Table 4.4 summarizes the comparative static results. The first column shows the qualitative effects related to stove adoption as derived by totally differentiating condition (4.11). The second and third columns show the qualitative effects related to stove efficiency, given adoption, as derived from conditions (4.9) and (4.10) for households with decreasing absolute risk aversion (DARA) and for households with increasing absolute risk aversion (IARA).[15] (See Silberberg 1989. Our appendix 4B provides the proofs.) This qualitative evidence anticipates the best household and community targets for development programs that feature improved stoves.

Table 4.4. Anticipated Effects of Exogenous Parameters on the Probability of Improved Stove Adoption and the Efficient Level of Stove Use

Parameter	P(adoption)	Efficient level of use S	
		DARA[a]	IARA[b]
Change in fuelwood price, p_f	+	+	+
Change in price of other (substitute) Consumption goods, p_o	+	+	+
Change in wage, w	0	+/−	−
Decrease in fuelwood consumption [mean decrease in $\alpha(e,S)$]	+	+	+
Variance in fuelwood consumption [mean preserving spread in $\alpha(e,S)$]	+	+/−	+/−
Change in household profit, $\pi(.)$	0	+/−	−
Change in exogenous income, M	0	+/−	−

[a] decreasing absolute risk aversion
[b] increasing absolute risk aversion

All price increases (p_f or p_o) improve the likelihood of adoption because their direct and positive substitution effects outweigh their indirect income effects. Price increases also induce increases in stove efficiency. We expect that, as prices rise, households that are decreasingly risk averse (DARA) will invest more in stove efficiency than households that are increasingly risk averse (IARA).

The greater the anticipated mean fuel reduction due to use of an improved stove, the greater the likelihood of adoption, and of more efficient use as well. A change in the variance in expected fuel reduction, however, will be perceived differently by households with different attitudes toward risk. While risk preferring households always prefer an increase in the variance of expected fuel reduction, the qualitative effects on investments in improved stoves in these households are ambiguous.

Greater income (household profits or exogenous income) has no effect on adoption. It has an uncertain effect on stove efficiency because risk aversion can change over income levels. If, as generally thought, higher income households are less risk averse (i.e., they exhibit DARA), then the income effect can dominate and higher income households may be willing to use a greater share of their income for fuel—or they may be willing to invest more capital in stove efficiency.

In sum, these qualitative results argue that prices and technology-induced fuelwood reduction are the important factors in determining adoption. Predicting efficient stove use is a more complex, and less certain, issue. It depends on the unobservable household aversion to a risky investment. For either decision, we may still anticipate that unexplained household characteristics also can be important indicators of both adoption and efficiency.

Empirical Results

Our survey of 99 households from Nepal's hills provides the data to investigate the behavioral postulates derived from conditions (4.9)–(4.11). The survey includes observations on fuelwood; agricultural residues consumed as a substitute fuel, profits from household production, and exogenous (business and service) income; various demographic characteristics; and qualitative variables identifying whether the household adopted an improved stove and the efficiency of stove use. The survey also includes observations on the reduction in fuelwood use after a stove was adopted. All households in the sample burned fuelwood.

Households confront the stove adoption and efficient use decisions simultaneously, according to conditions (4.9)–(4.11). These decisions can be expressed in a simple probit framework. For the first decision:

$$P(\text{adopt}) = \Gamma(Z); \quad P(\text{do not adopt}) = 1 - \Gamma(Z) \qquad (14.12)$$

where, from condition (4.11), Z is a vector that includes fuelwood price, household demographic characteristics, and other variables that affect the formation of expectations about the reduction in fuelwood consumption. $\Gamma(Z)$ is the cumulative density function of a normal random variable. For simplicity in the second decision [from conditions (4.9) and (4.10)], consider two improved stove types, with efficiency levels $S_1 < S_2$. Then

$$P(S_1 \text{ given } S > 0) = \Gamma(Z \text{ given adopt});$$
$$(14.13)$$

$P(S_2 \text{ given } S>0) = 1 - \Gamma(Z \text{ given adopt})$

Table 4.5 presents the probit estimates for (D4.1)–(D4.2).[16] These results are generally consistent with the theoretical expectations from table 4.4. The survey does not include observations on wages or on the households' variances in expected fuel reduction. Therefore, we cannot draw conclusions about the wage rate or the mean preserving spread. Furthermore, the survey includes no observations of the prices of other consumption goods. It does, however, include observations on the consumption of agricultural residues, a substitute for fuelwood. If residue demand is downward sloping, then consumption decreases when residue prices increase and an agricultural residue proxy for p_o should have the opposite sign from that anticipated for p_o in table 4.4.[17]

Table 4.5. Probit Estimates of Decisions 1 (Adopt) and 2 (Efficient Use)[a]

Independent variables	D1 adoption	D2 efficiency
Constant	-2.69* (-3.20)	-1.52* (-3.39)
Fuelwood price, p_f	0.0641** (1.94)	0.0164 (0.67)
Agricultural residues (a negative proxy for p_o)	-0.293* (-2.57)	-0.110** (-1.86)
Fuelwood reduction, $\alpha(e,S)$	0.143* (1.99)	0.124* (2.27)
Profit from household production, π	-0.000567 (-0.22)	-0.00159 (-0.77)
Exogenous (non-agricultural) income, M	0.00859* (2.51)	0.00575** (2.15)
Brahmin	1.072** (1.93)	0.663* (2.00)
Likelihood ratio test statistic	24.3	15.2
Percentage of correct predictions	90	76

[a] Numbers in parentheses are asymptotic t ratios
* Significant at the 0.01 level, ** significant at the 0.05 level

The adoption equation prices have the anticipated signs and both are significant. Fuelwood reduction is positive and significant, as anticipated. The income terms (π and M) are small and one is insignificantly different from zero, as it should be. The efficiency equation prices both have the anticipated signs and one is significant. Fuelwood reduction is positive and significant, as anticipated. The significant positive coefficient on exogenous income (and the insignificant coefficient on profits) suggest decreasing absolute risk aversion—a reasonable expectation for most populations. Brahmins are the highest Hindu social caste. They are generally wealthier and better educated. As a group, they are probably better informed about new opportunities and they probably have more capital to use in risky activities than the average household in the general population. A dummy variable identifying Brahmin households performs well in both equations.[18]

The test statistics for both equations are satisfying. The adoption specification predicts ninety percent of all observations correctly. The efficiency specification predicts 76 percent of all observations correctly.

Conclusions

This chapter examines household fuelwood production and consumption and the adoption of improved stoves, a technological substitute for fuelwood. The important finding of the production analysis is that the role of women and children in fuelwood collection is important, but not unique. In two districts in east-central Nepal we found that women and children are the significant collectors for those households that rely on community forests for their fuelwood. Adult males, hired labor, children, and agricultural capital are also significant inputs to fuelwood production for households that produce large shares of their fuelwood on their own private agricultural lands. Indeed, the contribution to fuelwood production on private land in our sample is significantly greater for adult males than for adult females. Apparently, as deforestation occurs on the common forestlands and fuelwood becomes scarcer, households begin to rely on their own agricultural lands to grow trees, and the role of men increases in the production of fuelwood.

Our consumption evidence tells a consistent story: Both fuelwood and combustible crop residues tend to be inferior goods. Fuelwood is inferior for higher income households and in Sindhupalchok where there is a more active and higher priced fuelwood market. Residues are inferior for the general population, but not for lower income households. These observations are true for farm and non-farm households alike. Our proxy for a cross-price effect argues that households substitute residues for fuelwood only reluctantly. The comparative coefficients on improved stoves in the fuelwood and residue demand functions suggest that residues are a less preferable material and that most households substitute fuelwood for residues whenever they can.

These consumption observations should have favorable implications for Nepal's environment as household incomes grow. If low income households grow to behave like our higher income households as the economy grows, then more households will convert from residues to fuelwood. The additional residues remaining on the ground will help sustain soil productivity. As household incomes increase further, agricultural households may grow more of their own fuelwood and non-agricultural households will convert to substitute fuels and fuel technologies. This all suggests that forestry development activities may be most important for Nepal in the period before development proceeds very far, and before traditional demands on the forest environment change.

Improved stoves can be an important fuel conserving, and environment protecting, technology. Not all households adopt new stove technologies and not all adopters operate their stoves at the same level of efficiency. Thus, the adoption and efficiency decisions imply two levels of potential fuel saving. Policy makers must consider both when evaluating the reduction in deforestation due to introducing improved stoves.

Furthermore, both adoption and efficient use are uncertain events. Neither is a sure thing. Both are dependent on the relevant prices, uncertain fuel savings, and household risk preferences. Households in higher price regions, households expecting greater fuelwood savings, and households with demographic characteristics indicating greater information and more risk capital will adopt the new technology before their less-well-endowed neighbors. Among these households, our evidence suggests that those with greater exogenous incomes will use the new technology more efficiently.

Appendix 4A. Input-Output Separability and Joint Production

This appendix justifies the allocations of inputs for the preferred regressions in table 4.1. For the Sindhupalchok households in our survey, a large share of fuelwood production occurs on each household's private land. Our male and female labor input data refer strictly to fuelwood production. Nevertheless, these inputs, as well as our capital inputs, may also contribute to crop and residue production. Crops and residues are joint products of these inputs and their production is not restrictive of fuelwood production. Therefore, there are no restrictions on our input choices for Sindhupalchok fuelwood production.

Fuelwood production in Kavreplanchok is different. We must confirm that adult male labor, hired labor, and capital are inputs in the joint production of crops and residues, leaving adult female labor and child labor for fuelwood collection from community forests. Confirmation requires a) evidence that adult male labor, hired labor, and capital are *not inputs in fuelwood production*, b) acceptable specification of the joint crop-residue cost function for Kavreplanchok using these three inputs, and c) evidence, from the parameters of this cost function, of input-output separability.

The second (private) regression for Kavreplanchok in table 4.1 adds the adult male, hired labor, and capital inputs to the first (commons) regression, without improving the fuelwood production equation. A separate specification of the commons regression, adding adult male labor, is also no improvement. The adult male coefficient is insignificant and the equation's statistical fit is poorer. Therefore, we can reject an hypothesis that fuelwood in Kavreplanchok is generally produced with these three inputs.

Table 4.A1 shows the generalized translog multi-output (crop-residue) cost function for Kavreplanchok. All observations are complete and no data is censored. The dependant variable is total agricultural production expenditures. The absence of measures of crop production, wages, and unit capital costs causes us to use proxies. The signs on the four direct coefficients are satisfying and seven of eleven coefficients are statistically significant. (Signs on the interactive terms are difficult to predict.)

The statistical significance of the food crop and wage proxies, and of three interactive terms indicate joint production. The equation F test shows that this translog cost function works well. Therefore, we cannot reject the hypothesis that crops and residues are jointly produced with the combination of adult male and hired labor and agricultural capital.

Finally, a procedure proposed by Hall (1979) confirms the input-output separability necessary to estimate a cost function of the translog form. This procedure specifies restrictions on the cross-product terms in the translog cost function that confirm jointness. Likelihood ratio tests do not reject Hall's restrictions on our cost function. Therefore, they support our contention of jointness.

Of course, none of this evidence denies that women work in the fields. A better survey for our purposes would have separated female employment into two categories: fuelwood collection and work on field crops. This is something to consider in the design of future fuelwood research.

Appendix 4B

Adoption

The qualitative effects of adoption can be derived by totally differentiating condition (4.11). From condition (4.11) and the assumption that $U_w^1(.) \neq 0$, the condition to adopt an improved stove simplifies to

Table 4.A1. Translog Multi-Output Cost Function Estimates for Food Crop and Agricultural Residue Production in Kavreplanchok[a]

Constant	3.36**
	(1.71)
Log residue output	-40.99**
	(-1.65)
Log agricultural income (proxy for food crop output)	0.69**
	(1.64)
Log collection time (negative proxy for wage)	-3.37
	(-1.50)
Log bovines (negative proxy for unit capital cost)	-51.84**
	(-1.78)
Log (AB)	-5.33**
	(-1.81)
Log (AC)	-0.093
	(-0.05)
Log (AD)	46.02**
	(1.71)
Log (BC)	0.0073
	(-0.02)
Log (BD)	5.26**
	(1.71)
Log (CD)	1.64
	(0.49)
Equation F	5.16**
Log of the likelihood function	109
Degrees of freedom	39

[a] Numbers in parentheses are asymptotic t ratios
*Significant at the 0.01 level, **significant at the 0.05 level

$$E\frac{\partial \alpha}{\partial S}(e,0) - \frac{1}{p_f} = 0 \qquad (4.B1)$$

$$\rightarrow S > 0 \quad \text{iff} \quad E\frac{\partial \alpha}{\partial S}(e,0) < \frac{1}{p_f}$$

It is apparent that, for a given $\partial\alpha(e,0)/\partial S$, an increase in p_f implies an increase in the probability that $S>0$. Similarly, for a given p_f, an increase in $\partial\alpha(e,0)/\partial S$ increases the probability that $S>0$. Because they do not appear in condition (4.B1), neither p_o, w, π, nor M have any effect on adoption.

Efficiency

Using $\alpha(e,S)=\alpha Se$, the second derivatives for conditions (4.9) and (4.10) are:

$$V_{ss} = E[U^1_{ww}(.)(\alpha e - 1/p_f)^2] \qquad (4.B2)$$

$$V_{oo} = E[U^1_{ww}(.)(-p_o/p_f)^2 + U^2_{oo}(.)] \qquad (4.B3)$$

$$V_{sM} = V_{s\pi} = E[U^1_{ww}(.)(\alpha e - 1/p_f)(1/p_f)] \qquad (4.B4)$$

Differentiating conditions (4.9) and (4.10) with respect to each exogenous variable yields:

$$V_{sp_f} = E[-U^1_w(.)(1/p_f^2) + U^1_{ww}(.)W(\alpha,S)/p_f^2] \qquad (4.B5)$$

$$V_{sp_o} = E[U^1_{ww}(.)(\alpha e - 1/p_f)(-Y_o/p_f)] \qquad (4.B6)$$

$$V_{sw} = E[U^1_{ww}(.)(\alpha e - 1/p_f)(t-R)/p_f] \qquad (4.B7)$$

$$V_{Se} = E[U^1_{ww}(.)(\alpha e - 1/p_f)(\alpha S) + U^1_w(.)(\alpha)] \qquad (4.B8)$$

$$V_{oM} = V_{o\pi} = E[U^1_{ww}(.)(-p_o/p_f^2)] \qquad (4.B9)$$

$$V_{op_f} = E[U^1_{ww}(.)(-p_o/p_f)(W(\alpha,S)'p_f^2) + U^1_w(.)(p_o/p_f^2)] \quad (4.B10)$$

$$V_{op_o} = E[U^1_{ww}(.)(-Y_o/p_f) - U^1_w(.)(1/p_f)] \quad (4.B11)$$

$$V_{ow} = E[U^1_{ww}(.)(-p_o/p_f)(t-R)] \quad (4.B12)$$

$$V_{oe} = E[U^1_{ww}(.)(-p_o/p_f)(\alpha S)] \quad (4.B13)$$

where $W(\alpha,S)$ is defined in the body of the chapter.

Standard comparative statics will show that for any exogenous variable z and where $J = V_{ss}V_{oo} - (V_{so}V_{os})^{1/2} \leq 0$.

$$\frac{\partial S}{\partial z} = \frac{V_{oz}V_{so} - V_{sz}V_{oo}}{J}$$

$$\frac{\partial f_o}{\partial z} = \frac{V_{sz}V_{os} - V_{oz}V_{ss}}{J}$$

The results in table 4.4 can be derived by replacing z with M, π, p_f, p_o, w, and e and using the fact that the coefficient of absolute risk aversion is $\rho(W) = -U_w(.)/U_{ww}(.)$. The complete proofs are lengthy. They can be obtained from the authors.

NOTES

1. Agricultural Projects Service Centre, Kathmandu, Nepal; Forestry Department, VPISU, Blacksburg, VA; and Centre for International Forestry Research; respectively. Keshav Kanel and Arun Malik provided helpful comments on early drafts of this chapter. The chapter benefits from critical comments made during seminars at the Chinese Academy of Forestry and with a group of Indian agricultural and forest economists organized by Winrock International and ICRISAT in September and October 1991. Earlier versions of this chapter appeared in two parts as "Joint production and consumption in traditional households," *Journal of Development Studies* 30(1):206–225; and "The adoption of consumption technologies under uncertainty," *Journal of Economic Development* 17(2):93–105.

2. An average Nepali district is approximately twenty miles square.

3. The best way to estimate these production decisions would be to specify complete input demands for labor and capital. Lack of reliable wage data will force us to estimate more straightforward production functions.

4. This implies that heteroskedasticity in the cross-sectional data is a greater problem than the potential variation from more accurate non-linear functional forms. We will assume that all heteroskedasticity is of the form $(a+b_i Z_i)^2$ where a and the b_i are variance parameters and the Z_i are a subset of independent variables in the tobit model.

5. Community forests may range from well-managed commons to open access resources. This raises an important general resource management issue, but an issue that is irrelevant for our level of analysis so long as households can collect from some existing common forestlands accessible to the district.

6. We also respecified the commons production regression, adding adult male labor. The adult male coefficient is insignificant in this regression in both districts. Furthermore, the addition of adult male labor fails to improve the statistical fit of the commons production regression in either district. These results support the usual contention that women are the primary fuelwood collectors from common forestlands.

7. The coefficient on adult male labor in Sindhupalchok is significantly greater (at the 0.01 level) than the coefficients on either adult female or hired female labor.

8. We also ran a switching regression using the full 99 observations. This regression separates households that purchase (but do not collect) from households that only collect fuelwood. Of course, those households that purchase fuelwood each report an observation on fuelwood price. The price coefficient in this regression has the expected negative sign and is significant at the 0.05 level. The price elasticity (evaluated at the mean) for these few households is −0.49. These households are generally not so poor. Apparently their personal labor opportunity costs are greater than the implicit wage for fuelwood collection. The remaining specification of the switching regression is similar to the regressions in table 4.3, except that its independent variables are less significant. The switching regression adj $R^2=0.85$, $F=17.1$.

9. Neither theory nor our empirical observations support an argument for income distinctions between households that own or do not own improved stoves. Theory does suggest that higher income households are more likely adopters of improved stoves (we address this issue in the next section), but the likelihood of adoption does not enter the regressions in table 4.2. The insignificance of the low income coefficient on improved stoves reflects the fact that fewer low income households possess improved stoves.

10. Once more, 25 observations on fuelwood price, most of them from higher income households and more of them from Sindhupalchok, are an insufficient number for statistical reliability. The alternative independent variable is collection time for the last unit of fuelwood

11. Indeed, a glance at the individual survey observations seems to support this hypothesis. We have no observations on land ownership. Agricultural income, however, may be a reasonable proxy. Agricultural income ranges from zero to 43,000 rupees for households in Sindhupalchok, residue consumption ranges from zero to 220 kg, and fuelwood collection time ranges up to eight hours per headload. Of thirty households earning less than 12,000 rupees of agricultural income, seven have

supplemental business or service incomes, creating total incomes above the median. Of the remaining 23 low income agricultural households, nineteen spend five or more hours collecting their marginal headloads of fuelwood and eleven of those nineteen burn less than forty kg of residues per month.

    12. Improved stoves are all very similar in this part of Nepal. S is a continuous variable because some households expend more resources to use similar stoves more efficiently.

    13. Fabella (1989) classifies and provides examples of the un-certainties normally encountered in household production.

    14. In practice, the household can adjust its level of stove expendi-tures continuously. Our data, however, indicate only two levels of stove efficiency. Therefore, our empirical analysis abstracts from the continuous case and explains efficiency as a discrete variable. Household choice between two discrete levels of efficiency can still be represented with expected utilities. [For example, given that the household adopts a stove, the choice between two stove types ($S_1 > S_2$) is described by the expression: Choose S_2 if and only if $EV(p,I|S=S_2)] > EV(p,I|S=S_1)$—where $V(p,I|\cdot)$ is defined as in eq. (4.7), but is conditional on the level of stove efficiency.]

    15. Using the definition of the Arrow-Pratt coefficients of risk aversion, $A(W) = -U_{ww}/U_w$, DARA implies that $dA(W)/dW < 0$. IARA implies $dA(W)/dW > 0$. These restrictions are useful in assessing the comparative static results in table 4.4. The assumption of separability in the utility function does not effect the interpretation of the Arrow-Pratt coefficient of risk aversion.

    16. We assume that errors in conditions (4.9)–(4.11) are symmetric and uncorrelated. The validity of these assumptions can be tested using bivariate probit methods—when larger data sets become available.

    17. The functions in table 4.3 already confirm that residue demand conforms with general economic expectations.

    18. We examined, and rejected, econometric specifications including a dummy variable for Newars, the ethnic group that uses additional fuel to brew a traditional alcoholic beverage. We also rejected specifications including years of school and family size.

    The two survey districts have substantially different forest endowments. Within each district, farm houses are more likely to be of individual construction than houses in villages. Therefore, farm houses may require more heating. They are also closer to the sources of fuel. Nevertheless, specifications that separate the 99 observations by district, primary income source (agricultural or business-service), or income level do not improve on the results for the full sample.

REFERENCES

Fabella, R. V. 1959. "Separability and risk in the static household production model." *Southern Economics Journal* 55(4): 954–62.

Hall, R. E. 1973. "The specification of technology with several kinds of output." *Journal of Political Economy*. 81(4):878–92.

Madalla, G. S. 1983. *Limited Dependent and Qualitative Variables in Econometrics.* Cambridge: Cambridge University Press.

Silberberg, E. 1989. *The Structure of Economics: a Mathematical Analysis.* 2d ed. New York: McGraw-Hill Book Co.

Singh, I., L. Squire, and J. Strauss. 1986. "The basic model: theory, empirical results, and policy conclusions." In I. Singh, L. Squire, and J. Strauss, eds., *Agricultural Household Models.* Baltimore: Johns Hopkins University Press, pp. 39–69.

CHAPTER 5

Innovation and Adoption in Pakistan's Northwest Frontier Province

Mohammad Rafiq, Gregory S. Amacher, and William F. Hyde[1]

Consumption technologies like improved stoves are important, but production technologies like new species, new nursery stock, or new planting techniques are more common forestry activities. Specialized exotic species like mangos from Southeast Asia, lucaena from the Caribbean, and eucalypts from Australia, have been introduced around the world. Faster growing, multiple purpose trees like lucaena, populus, and acacia are central to numerous environmental and rural development projects. Even planting and managing indigenous species is a new technology where natural regeneration has traditionally satisfied all local demands.

Consumption technologies like improved stoves have both substitution and wealth effects on household consumption. The substitution effect decreases fuelwood consumption. The wealth effect increases fuelwood consumption, although it is unlikely that this increase is sufficient to offset the fuel conserving substitution effect.

For both consumption and production technologies, households must balance the uncertain gains from the new technology with the certain costs of its inputs; for example, the costs of purchasing stoves or a new seed variety. New *production* technologies, however, create an analytically more complex economic problem. They increase forest-based production and release household labor from forest-based activities for use in other productive activities. The general increase in overall household production (forest-based production plus other production created with released household labor) enables increases in general household consumption and, therefore, in general household welfare.

Agriculture's experience with innovation and the adoption of production technologies [see Feder et al. (1985) and Ruttan (1977)] has much to teach us. Agriculture, however, is usually concerned with the diffusion of a new or improved input for producing a known output. Forestry often contends with the additional complexity of a new input-output mix (*e.g.*, new seedlings that require additional periodic management and eventually produce a tree that provides the old tree outputs, but also adds new outputs like fodder or nitrogen fixation). The additional uncertainty attached to the new input-output mix is one reason why rural households may be slower to adopt new forestry technologies. The policy implication of slower adoption is that new forestry activities will require longer periods of external advice, demonstration, and training before the new activities become self-sustaining.

In order to organize our understanding of this problem, we will sketch a model of the representative household's decision to adopt a new production technology. (Our nomenclature is comparable with the stove model in the previous chapter.) This model will illustrate that greater household incomes, larger household capital and labor stocks, and more secure land tenure are useful indicators of those households that are most likely to adopt new forest production technologies. Better market access and higher product prices probably, although not certainly, indicate communities that are more likely to adopt new forest production technologies.

Experience from six villages in Pakistan's Northwest Frontier Province illustrates the model's conceptual findings. The six villages were all eligible for forestry assistance from the Forestry Planning and Development Project, an 84 month (1984–91) joint effort of the Government of Pakistan and the US Agency for International Development, with a capital outlay of Rs 536 million (1991 US $30.6 million). Interviews with citizens from the six villages support our expectations of adopting households. Furthermore, the interviews remind us to go a step further and consider the characteristics of institutional delivery of the new technology, as well as the characteristics of household demand. In particular, we will find that the villagers emphasize the importance of the personal character, rather than the professional expertise, of extension foresters who are responsible for transferring new technologies to local households.

The Conceptual Model

Consider a representative household's decision to obtain and plant new seedlings—for the first time. The household maximizes a utility function with arguments for the consumption of fuelwood and other goods. Both fuelwood production and the household budget constrain consumption. The household obtains fuelwood from community forests. Its problem is to select its optimal use of productive inputs (therefore its optimal investment in new seedlings) and its optimal consumption pattern. The new seedlings may be expensive to purchase. Or they may be free but obtaining them from the nursery may require a commitment of household labor. Planting and management may also require commitments of household labor and capital not previously devoted to tree growing.

The complete household problem is

$$V(p,I;C) = \max EU(Y_j;C) \quad j=f,o$$
$$\text{s.t.} - \Sigma_j p_j Y_j + M + \tilde{\pi} = 0 \quad j \neq f$$
$$\text{where } \tilde{\pi} = p_f Q_f - rL \quad (5.1)$$
$$Q_f = f(\mu, L) = f(\mu * L) \quad f' \geq 0, f'' \leq 0$$
$$Y_f, Q_f, L \geq 0$$

where maximization is over household consumption of forest-produced goods Y_f and other consumer goods Y_o, and household use of productive inputs L. V(.) is the household's indirect utility function. p is a vector of known market prices, I is household income, and C is a vector of exogenous characteristics that influence household behavior. U(.) is household utility. M is exogenous income, which includes income from other (non-forest) household agricultural activities, wages from the off-farm hire of family labor, and non-farm income. $\tilde{\pi}$ is the household return from forestry activities, r is the cost of tree planting, and L is a proxy for the combined use of capital, labor, and land in the new tree planting technology. Household income is the sum of exogenous income and the household return from forestry. It equals the summed household uses of consumer goods times their prices.

Forest-based production Q_f has a random component μ defined by the probability density function $g(\mu)$ over the interval $(0,\infty)$. For illustrative purposes, we will assume that forest-based production is linear and that the random component enters as multiplicative uncertainty. This makes $\tilde{\pi}$ a random variable dependent on the level of planting effort L. Expected household profits from forestry and the expected level of physical forest productivity both increase with increases in planting effort.

Production uncertainty makes this model more complex than the stove adoption model in chapter 4. Random production introduces randomness into consumption through the household budget. Therefore, the household's choice of its level of planting effort is no longer separable from the household consumption decision.

Nevertheless, it is possible to simplify the model while maintaining our search for the indicators of the household choice under uncertainty. One approach would be to hold the consumption of other goods Y_o fixed and to assume the household collects all of its own consumption of forest products Y_f. That is, $Y_f = Q_f$.

90   Economics of Forestry and Rural Development

The household's decision becomes

$$\max_L \ E[U(I^*)] \tag{5.2}$$

$$\text{where } I^* = \frac{\tilde{\pi}}{p_f} + \frac{M}{p_f} - p_o \frac{Y_o}{p_f}$$

and the expectation is taken over μ. The household now faces a random household budget I^*.

The first order condition can be found using the chain rule.

$$EU'(I^*)[f'(\mu,L)\mu - r/p_f] = 0 \tag{5.3}$$

Alternatively,

$$E[U'(I^*)f'(\mu,L)\mu] = \frac{r}{p_f} E U'(I^*) \tag{5.3a}$$

Totally differentiating eq. (5.3a) and using the implicit function theorem

$$\frac{dL}{dI^*} = -\frac{EU''(I^*)[f'(\mu,L)\mu - \frac{r}{p_f}]}{EU'(I^*)f''(\mu,L)\mu^2 + EU''(I^*)[f'(\mu,L)\mu - \frac{r}{p_f}]^2} \tag{5.4}$$

Eq. (5.4) represents the household's willingness to invest in increments of L as the components of household income change.

Conventional wisdom anticipates that most households are risk averse, but that they become decreasingly risk averse as household wealth increases. We can use the Arrow-Pratt measure of risk aversion and the definition of decreasing absolute risk aversion (DARA), to show that eq. (5.4) is positive. (See Silberberg 1989. Appendix 5A contains the proof.) This indicates that, in the conventional case, as household income increases, the household's willingness to invest in an uncertain new tree planting technology also increases.

Table 5.1 examines how the household's willingness to invest is affected by the exogenous variables in the household budget. The variables in the table are more narrowly defined than the variables in our formal mathematical model. Therefore, the symbol identified with each variable in the table is the symbol from our equations that *includes* this variable (and perhaps other variables in the table as well).

Table 5.1. Exogenous Determinants of Adoption of a New Planting Technology—Assuming Decreasing Absolute Household Risk Aversion (DARA)

Variable	Efficient level of planting technology
Change in exogenous income, M	+
Change in agricultural income, M	+
Change in forest-based income, π	+
Change in household endowment of labor, L	+
Change in household endowment of capital and land, L	
Change in price of forest products, p_f	0 or +
Change in price of other consumption goods, p_o	−
Change in variance of forest-based production [mean preserving spread in μ]	−/+*

* Introducing a mean preserving spread is more difficult in this case than for the consumption technologies discussed in chapter 4. Uncertainty from production technologies like planting appears in both the production function and the budget constraint. This implies non-separable consumption and production decisions. Nonseparability makes the qualitative effects of a mean preserving change in μ ambiguous—although the usual assumption is that its sign is negative under DARA.

Consider, more generally, the classes of households that this analysis predicts as most likely to invest in uncertain new production technologies. First, higher income households are more likely to invest, regardless of their income source. Second, households with larger endowments of the various productive factors; land, labor, and capital; are more likely to invest. Additional endowments of either income or the household's various productive factors suggest greater household flexibility in both production and consumption decisions and greater likelihood of achieving any perceived level of absolute household welfare. Furthermore, education, access to information, and forestry technical assistance all improve the quality of household labor. Quality improvements effectively increase the factor endowment and improve household willingness to try a risky new technology.

Third, households in the neighborhood of improved roads or otherwise improved market access may also be inclined to invest more in uncertain new technologies. Market access (therefore, easier exchange) increases the external demand for local producer goods, forest products in our case. This increases forest-based household income, but it also increases the cost of local domestic consumption of the same goods. The net effect on the household budgets and household willingness to accept risk tends to be positive because households would not engage in trade if the net effect were negative. New markets can only leave households as well off or wealthier than they were in the absence of the new markets.

Improved market access generally decreases the local price of consumption goods otherwise less available to local households. Decreasing local consumer prices increase the relative value of the household budget and increase the household's willingness to accept risk.

Finally, consider the factors that might decrease the variance in expected forest-based production. First, consider land tenure and the local political institutions. Where the land used for forestry is neither private nor a well-managed commons, then we might anticipate that the variance on the expected productivity of any new planting technology is great. Households will be most uncertain of their abilities to capture any productivity gain from the new technology. Therefore, we anticipate a reluctance to invest. Village elders, with greater control over both the fragile rights to the land and its eventual productivity, will be more likely than the poorest households to invest in new planting technologies on this land. The poorest households will be unwilling to invest any of their own scarcer personal resource endowments. They will be pleased for others (like external development agencies) to invest, but these poorest households will be even more interested (than the wealthier and more powerful village elders) in more certain and shorter-term gains. Therefore, we might expect the poorest households to be more interested in agricultural production than in longer-term forestry activities. For the same reason, we might expect the poorest households to be more likely to crop new trees before they become mature. Where tenure is insecure, yet where they must accept forestry investments, the poorest households will be more interested in the more certain early yields from fast growing tree species, even where slower growing species might yield greater long-term production.

Both forestry technical assistance and prior success by neighboring adopters might also affect the variance in expected productivity from the new planting technology. Both increase local knowledge, decrease the variance on household expectations, and make adoption more likely. We might anticipate that both will encourage better endowed households to adopt first. Some poorer endowed households will follow as they gain confidence from the earlier successes of their better-off neighbors.

Background and Empirical Evidence

Pakistan is the tenth most populous and 24th poorest country in the world. Its GNP per capita was only US $320 in the mid-1980s. Its population is 76 percent rural, but the countryside is semi-arid and only 26 percent of the land is cultivable. The five percent of the country that is forested is mostly in low productivity semi-arid or alpine zones or in watershed protection areas. Yet fuelwood—supplemented with animal wastes and crop residues—is the country's primary energy source. Ninety percent of the rural population uses these fuels, and wood provides one-half of Pakistan's household energy. Energy consumption is 0.20 m^3 per capita per annum, the lowest in South Asia (US Department of State 1983), perhaps because of the scarcity of better forest resources.

Real fuelwood prices have increased 200 percent in a decade and forest inventories continue to decline. One response, however, is increasing private wood production. Public and community forests now provide only twenty percent of Pakistan's sawtimber and 25 percent of its fuelwood. The remainder comes from scattered privately owned trees, and the adoption of new seedlings and other tree planting on private lands has been an issue of great interest.

Now consider our propositions from earlier in this chapter for the region of Kohat, an isolated district of Pakistan's Northwest Frontier Province. The Northwest Frontier Province is semi-arid to sub-humid and mountainous. It is located high on the Indus River where the Karakoram, the Hindu Kush, and the Himalayan geologic plates come together. Kohat itself is adjacent to the famous Khyber Pass to Afghanistan.

The district population of 509,000 is supplemented—and stressed—by an Afghan population of 208,000 refugees that generally camps near the remaining forest and draws heavily upon it (Government of NWFP 1985, and Office of the District Administrator, Afghan Refugees, Kohat). The local Pakistani population exports male labor to the Middle East, from which it obtains large inflows of wealth. Women, on the other hand, generally remain in the village and seldom work outside the home. We will see that each of these factors; prices, refugees, external wealth, local labor supply; may influence the household response to new forestry technologies.

Land, rifles, and brick homes are symbols of wealth, but landholdings average only 11.85 acres per household, 2.75 of which are cultivable and one-third of that are irrigated. State forests comprise only a declining and deteriorating 1.8 percent of the district's land area (Khan and Khan 1986). These are all good reasons why many farmers have begun growing trees for their own domestic wood consumption, usually near their homes and farm buildings, along the boundaries of their fields, and along irrigation channels.

Data

The Government of Pakistan introduced the Forestry Planning and Development Project in Kohat and several neighboring districts in 1983. The

Project's objective was to increase the indigenous wood energy supply, with specific emphasis on private farm lands. The Project encouraged farmers to plant new seedlings and to participate in other forest production activities, but it did not provide large subsidies to farmers to accomplish this objective. Therefore, the Project's impacts are an unbiased reflection of farmer preferences.

Our analysis examines one sub-area of the Project. It compares the endowments of six villages in Kohat, three of which chose to participate in the Forestry Planning and Development Project and three which chose not to participate, using secondary data from various government sources (Government of NWFP 1985, Khan and Khan 1986, and unpublished government records). These data are enriched with information from group interviews with 60 village elders, seven "tree-minded" farmers, and 72 additional heads of households.[2] The tree-minded farmers are farmers who were already engaged in raising trees. For them the dominant forestry objective was fuelwood, and five of seven tree-minded farmers grew fuelwood for market sale. Members of all three groups were asked to identify those factors supportive of or problematic for farm forestry. In addition, we asked seven Forestry Department professionals and seven agricultural extension agents of their own perceptions of the factors explaining why some farmers and some villages adopted the Project's forestry activities, yet others did not.

Results

Table 5.2 records our results. It compares roughly with table 5.1. For each variable in table 5.1, table 5.2 identifies a physical proxy and reports its average level for a) three villages that chose to participate in the Forestry Planning and Development Project and b) three neighboring villages that chose not to participate. Our expectation is that each positive sign in table 5.1 will be matched by a larger average household endowment for participating villages than for non-participating villages in table 5.2. We anticipate that the positive sign on forest products price and the negative sign on the prices of other consumption goods in table 5.1 correspond with increased accessibility (decreased mileage) for participating villages in table 5.2.

Our sample of six villages is too small to assess statistical reliability. Nevertheless, our empirical evidence is most appealing. The direction of change satisfies expectations in every case except uncultivated open access wasteland per household. Yet the smaller amount of wasteland in participating villages actually reflects smaller forest stocks, which together with better market access and higher fuelwood prices, provides the motivation for participating villages to plant trees. Therefore, our wasteland evidence reinforces the incentive to participate and, indeed, supports our conclusions derived from other observations.

Table 5.2. Exogenous Determinants of Participation in Forestry Activities

Exogenous factor*		Participating villages	Non-participating villages
exogenous income, M	(+)	0.54	0.38
export laborers per household			
agricultural income, M	(+)	2.66	1.68
irrigated acreage per household			
forest-based income, π**	(+)	-	-
household endowment of labor, L			
number of persons per household	(+)	7.0	6.0
number of males per household***	(+)	3.3	2.8
distance to nursery in miles			
(technical assistance)	(-)	19.	42.
household endowment of capital and land, L			
number of livestock per household	(+)	31.15	11.52
agricultural land per household in acres	(+)	6.30	5.52
uncultivated open access wasteland in acres			
(including forestland)	(+)	5.74	12.48
market information, (reflects on prices, p)			
market access in miles	(-)	5.	21.
distance to the highway in miles	(-)	2.	13.

   *Parenthetical expressions reflect the direction of difference, from table 5.1, between participating and non-participating households.
   **We have no measure of forest-based income, although some farmers do sell fuel-wood and poles for construction.
   ***Data are unavailable for three of the six villages. These are the average numbers of males in the household for two participating and one non-participating village.

Evidence on household wealth and other suggestions of fuelwood price lend additional conviction. During our interviews, we made subjective estimates of household wealth based on housing construction, eating utensils, and furniture quality. On a subjective wealth scale from one to three, participating village households averaged 2.5, while non-participating village households averaged only 1.5. The apparent greater wealth of participating village households corresponds to our expectations about greater income and greater factor endowments.

Eighty-three percent of households in participating villages use fuelwood substitutes in contrast with only 46 percent of households in non-participating villages. Eighty-one percent of households in participating villages burn inferior crop residues while none of the households in non-participating villages burn crop residues. This fuel consumption pattern provides a further suggestion of higher fuelwood prices in participating villages.

Household Perceptions

Further evidence on household perceptions, regardless of residence in participating or non-participating villages, provides support a) for our contentions regarding the importance of secure land tenure and b) for examining the institutions for delivery of the new technology.

The tree minded-farmers avoided tenurial conflict either by planting close to their houses or by fencing around their young trees. Both strategies ensure better protection of their long-term forest investments. Tree planting conflicts tend to arise where tenure is less certain: along the boundaries of private lands and on common lands. In addition, trees planted along boundaries potentially compete with the crops of adjacent farmers for sunlight, water, and soil nutrients. Private trees planted on common lands are perceived by one's neighbors as an attempt to lay claim to those lands. Communal planting on the commons, on the other hand, is perceived as a means of laying community claim and, thereby, protecting the village from encroachment by the refugee population. Elders and heads of household in five of the six villages commented on this latter, "protection" role for trees planted on the commons.

The actions of the tree-minded farmers are evidence that secure land tenure is an important characteristic explaining adoption of new tree planting activities. The perceptions of the elders and heads of households is evidence that causality also runs the other direction, that planting is an activity that can improve the security of land tenure.

We have shown how adoption depends on household characteristics. Adoption also depends on the technology being delivered and its means of delivery. Our survey participants stressed these two points.

Table 5.3 compares the perceptions of heads of household, tree minded farmers, and village elders, with the expectations of Forestry Department extension specialists, regarding the true needs of the local villagers. Foresters believe the villagers need more assistance with tree planting and growing. The villagers themselves feel their greater needs are for assistance in obtaining seedlings and in better protecting their existing trees.

This difference in perceptions caused us to examine the villager-forester relationship more closely. We inquired of the same four groups; heads of household, tree minded farmers, village elders, and extension foresters; of the qualities most important for a good forester. Table 5.4 summarizes these responses. The major distinction is between the groups of villagers and the foresters. The villager groups uniformly report greater concern with the personal characteristics of a good forester, while foresters perceive that professional skills are more important. The average head of household, if anything, is more concerned than (better off) tree-minded farmers and village elders with the personal character of extension foresters. The last row of table 5.4 is consistent with the perceptions in table 5.3. Extension foresters feel they should provide technical assistance—but the villagers feel that forestry technical assistance is neither a service they desire nor an important skill for good foresters.

Table 5.3. Perceptions of the Contributions of Public Foresters to Farm Land Owners

potential contributors	survey group*			
	heads of household	tree-minded farmers	village elders	extension foresters
assistance in tree protection	50	86	–	29
seedlings	65	100	100	86
advice–technical assistance	31	14	–	71

* percentage reporting the perceived need

Table 5.4. Perceived Qualities of a Good Forester

perceived qualities	survey group*			
	heads of household	tree-minded farmers	village elders	extension foresters
good moral behavior	98	85	100	29
honesty	90	57	75	0
equal treatment for all	71	0	25	0
acquaintance with local conditions	8	14	25	57
professional competence	4	14	50	86
natural inclination toward extension	0	0	0	57
extension and training experience	0	0	0	86

* percentage reporting the perceived quality

Several factors may explain the different villager-forester perceptions: the secluded role of women in these villages and the villagers' desire to protect their women members from outsiders, the fact that most foresters do not come from these villages originally, and foresters' historic police-like role in protecting government forests from these same villagers. Certainly this historic role may have left some villagers with residual sense of mistrust, and some foresters with a residual sense of control that is no longer justified.

Conclusions

In any case, it is clear that the characteristics explaining adoption include choice of the right technology and of a responsive delivering agent—as well as a responsive consuming household willing to accept the risk attached to the uncertain gain from adoption. Responsive households probably have higher incomes and larger factor endowments. From a policy perspective, this means that the poorest of the poor are not a good target population for new forestry innovations. Rather, we must count on a) adoption by better-off farmers to demonstrate success and, thereby, decrease the perceived risk of their poorer neighbors, and b) extension foresters who share the personal trust of both poorer and not-so-poor farmers.

Appendix 5A. Proof that $\partial L/\partial I^* \geq 0$ when households exhibit decreasing absolute risk aversion (DARA)

Repeating eq. (5.4) from the body of the chapter:

$$dL/dI^* = - \frac{EU''(I^*)[f'(\mu,L)\mu - r/p_f]}{EU'(I^*)f''(\mu,L)\mu^2 + EU''(I^*)[f'(\mu,L)\mu - r/p_f]^2} \qquad (5.4)$$

Dividing the numerator and the denominator of eq. (5.4) by $U'(I^*)$,

$$dL/dI^* = - \frac{[EU''(I^*)/U'(I^*)][f'(\mu,L)\mu - r/p_f]}{Ef''(\mu,L)\mu^2 + [EU''(I^*)/U'(I^*)][f'(\mu,L)\mu - r/p_f]^2} \qquad (5A.1)$$

Eq. (5A.1) can be restated in terms of some function of the Arrow-Pratt coefficient of risk aversion, $\alpha(I^*) = U''(I^*)/U'(I^*)$. Under DARA, $d\alpha(I^*)/dI^* \geq 0$, and eq. (5A.1) becomes

$$dL/dI^* = - \frac{[E\alpha(I^*)[f'(\mu,L)\mu - r/p_f]}{Ef''(\mu,L)\mu^2 + E\alpha(I^*)[f'(\mu,L)\mu - r/p_f]^2} \quad (5A.2)$$

From the definition of an expectation and the probability density function $g(\mu)$, eq. (5A.2) is equivalent to

$$dL/dI^* = - \frac{[\int_\mu \alpha(I^*)dg(\mu)]^*B}{Ef''(\mu,L)\mu^2 + [\int_\mu \alpha(I^*)dg(\mu)]^*B^2} \quad (5A.3)$$

where $B = [f'(\mu,L)\mu - r/p_f]$, which is greater than one.

We can more easily interpret the sign of eq. (5A.3) by decomposing the integrals. First define a partition over the probability space of μ, μ_1 and μ_2, such that $I^*_1 \le I^*_2$. Eq. (5A.3) can then be rewritten as

$$dL/dI^* = - \frac{[\int_{\mu_1} \alpha_1(I^*)dg(\mu) + \int_{\mu_2} \alpha_2(I^*)dg(\mu)]}{Ef''(\mu,L)\mu^2/B + [\int_{\mu_1} \alpha_1(I^*)dg(\mu) + \int_{\mu_2} \alpha_2(I^*)dg(\mu)]^*B} \quad (5A.4)$$

If the household is risk neutral (or exhibits neither increasing nor decreasing absolute risk aversion), then the numerator of eq. (5A.4) is zero (U"=0), and $dL/dI^* = 0$.

If the household exhibits DARA (becomes less risk averse as household income increases), then the numerator and the latter two terms of the denominator are positive (by the effect of DARA on α). The first term on the denominator is negative because the production function is concave. We can anticipate that the first term in the denominator is greater than the latter two terms because a) the actual physical production function is unchanging with respect to household risk aversion, while b) household utility becomes less concave (more linear) with decreases in household risk aversion. This implies a positive sign on eq. (5A.4). The positive sign instructs us that anything that increases household income also increases the investment in forestry activities like planting new seedlings that the household perceives to yield uncertain rewards.

NOTES

1. International Union for the Conservation of Nature, Forestry Department, Virginia Tech, Blacksburg, VA and Centre for International Forestry Research, respectively. This chapter benefits from Michael Dove's advice on the underlying survey, and from critical comments made during a seminar at the Chinese Academy of Forestry in October 1991. An earlier version appeared as "Local adoption of new forest technologies," *World Development* 21(3):445–54.

2. The employment assignment of the senior author determined the choice of the six villages. The 60 elders are a self-identified 100 percent population. The 72 additional heads of household are a 100 percent population of those in the six villages on the days of the first interview and a follow up. Our choices of elders and heads of households are confirmed by similar choices for an anthropological study also based on these villages (Dove and Qureshi 1987).

REFERENCES

Dove, M. R., and J. Qureshi. 1987. "Farmer interest in planting trees and operating nurseries in the NWFP: district and town reports." Islamabad: Office of the Inspector General of Forests and Winrock International Technical Assistance Team.

Feder, G., R. Just, and D. Zilberman. 1985. "Adoption of agricultural innovations in developing countries: a survey." *Economic Development and Cultural Change* 33(2): 255–98.

Government of NWFP "Planning and Development Department Bureau of Statistics." 1985. NWFP Development Statistics. Peshawar: author.

Khan, H. M., and S. Khan. 1986. "Agricultural statistics of Northwest Frontier Province for 1985–86." Peshawar: NWFP Department of Agriculture.

Ruttan, V. 1977. "The green revolution: seven generalizations." *International Economic Review* 19:16–23.

Silberberg, E. 1989. *The Structure of Econometrics: a Mathematical Analysis.* 2d ed. New York: McGraw-Hill Book Co.

US Department of State, Agency for International Development. 1983. Project paper, "Pakistan - forestry planning and development" 391–481. Washington: author.

Part 4
Regional Supply and Demand

Our previous cases featured agricultural households themselves. The next two cases will examine more aggregate regional indicators of the potential impacts of social forestry activities. Data limitations generally force us to take our intuition on aggregate supply from broad measures of the physical forest stock, and to assume that aggregate demand is proportional to some measures of the market or some measure of household labor applied to collecting the forest product. The next two chapters will examine the importance of these limitations.

Chapter 6 examines deforestation around 31 cities in eight countries, including three in Southeast Asia. The rural location of the deforestation justifies the inclusion of this chapter in our book. The chapter begins with a physical measure of the standing forest stock, but modifies conclusions arising from this measure according to three characteristics (roads, terrain, and moisture) which substantially affect the volume available for harvest. The resulting improvements in prediction are greater than the improvements due to correct specifications of policy variables. This is a strong argument for incorporating these economic characteristics along with the physical characteristics that generally appear in all estimates of forest supply. The analysis in chapter 6 also reveals that national policy distinctions are important sources of variation in forest stocks. We will examine national policy impacts more closely in our final two sets of case studies.

Chapter 7 returns to household surveys to build insights to fuel-wood scarcity and the opportunities for new social forestry activities in 29 districts across Nepal's hill and tarai regions. This chapter too finds that resource stocks are poor indicators of resource scarcity. Some combination of market prices, and an indication of the changes that occur in household behavior in response to the resource price, will be better indicators of fuelwood scarcity and household responses to it. It is the relative scarcity of the forest resource and the local opportunity costs of new social forestry technologies (like seedlings, nurseries, and improved stoves) that explain the likely acceptance of a specific technology in any particular locality. Chapter 7 shows that fuelwood scarcity in Nepal may not be as great as it often has been portrayed, but that scarcity is critical for select populations in select regions of the country. Identifying these target populations and regions is a key prerequisite for successful social forestry activities. Higher local prices, and more elastic prices and substitution are good indicators of these populations and regions.

CHAPTER 6

Estimates of Economic Supply from Physical Measures of the Forest Stock: An Example from Eight Developing Countries

Kerry Krutilla, Jintao Xu, Douglas F. Barnes, and William F. Hyde[1]

The supply-side of natural resource commodity assessments generally begins with a physical estimate of the resource stock, of proven reserves. Often it goes no farther-wholly disregarding large differences in the economic content of the physical measure. The remaining analysis then typically features demand variables and the effects of policy interventions. Howe's (1979) natural resource economics textbook illustrates this problem for minerals. Minerals assessments may divide the stock into a 2x2 matrix of discovered and undiscovered, economic and uneconomic cells (known as the McKelvey Box), but even in this case the economic cells are generally defined by geological criteria alone. Fisheries has, perhaps, greater difficulty estimating the size of natural stocks, yet perhaps even greater reliance on purely biological appraisals of policy issues. Forestry relies on measures of the standing forest stock estimated without any regard for whether the stock is within or entirely outside the boundaries of reasonable economic accessibility.

Physical measures of the stocks are troublesome for policy analysis because policy is largely designed to affect the economic (and sometimes the marginally sub-economic) subset of all physical reserves. Furthermore, the analytical bias that comes with using physical measures changes over both time and space. For inter-temporal analyses of a single resource stock, the bias changes as local relative prices change. If the long time trends in resource commodity prices are generally downward while the standards for measuring resource stocks are unchanging, then the bias worsens with time.

For cross-sectional analyses of multiple stocks of the same resource the bias is different for each stock because the economically accessible share is different for stocks in different markets and for stocks subject to different measurement standards. The result can be large unexplained differences in market and policy performance across regions and countries. This suggests a particular problem for empirical assessments of global resource policy and for identifying the sources of global adjustments in resource stocks because those assessments tend to be cross-sectional. Global assessments of deforestation are an obvious current example.

Our objective in this chapter is to examine the importance of the difference between physical stocks and economic supplies. This requires an improvement in the usual specification of the supply side of the analysis. (Let us be very clear. It is not our intention to determine sources of policy effects, or even sources of deforestation. Our intention is to demonstrate the importance of accurate economic measures of the resource stock.) Therefore, we will begin with a simple model of the demands and supplies of wood fiber and forestland. We will specify supply as a function of terrain, road density, and distance from the market, as well as a physical measure of the standing resource stock. An empirical specification of our final reduced form, compared with an empirical specification of the same reduced form with standing forest stock as the only supply variable, will demonstrate the importance of economic revisions of the physical measure. The standing forests around 31 urban markets in eight African, Asian, and Latin American countries will provide the basis for our empirical test. We will find that corrections for economic supply explain more than forty percent of the variation in cross-sectional levels of the forest stock in these 31 regions.

Distinctions between countries explain another twenty percent of the variation in our cross-sectional levels of the forest stock. Once the stocks are measured and markets are taken into account, the remaining distinctions between countries might be summarized largely as differences in national policies and institutions. Therefore, this second result supports a growing body of literature (especially Deacon 1994, 1995; Ostrom 1995; and Alston et al. 1996) that identifies important links between political activities affecting ownership rights and global deforestation. Indeed, Deacon and Alston et al. encourage analysis like ours with their own cautions about stocks as measures of supply and distance as a limitation to economic activity at the frontier, respectively.

The contrast between the (more than) forty percent of all variation due to mismeasures of economic supply and the twenty percent due to policies and institutions supports two contentions: a) even modest improvements in measures of the forest stock will improve policy expectations and b) the standard physical measures of the stock can confuse and may even mislead the many evaluations of broad scale effects on the global forest.

Background

Since von Thunen (1875), it has been common economic knowledge that the share of total forest stock corresponding to useful supply declines as distance from the market center increases. Moreover, we also know that the forested regions closest to urban centers are generally under the greatest stress for deforestation. Yet the usual physical measures of standing forests ignore locational distinctions that would stress the importance of some stocks while removing others altogether from the base of economic and policy concern.[2]

Foresters rely on physical measures of standing forest stocks because these fit their biological orientation and because unchanging physical measures provide uniform points of reference over time (called "continuous forest inventories"). In the US, for example, official estimates of "commercial" forest stocks begin with an arbitrary estimate of forest land and then measure the stock within this land area according to very specific biological criteria (Powell et al. 1993). Forestland is land capable of growing twenty cubic feet per annum of native forest. One common official measure of the stock within this land area is standing timber 11 1/2 inches in diameter at breast height (4 1/2 feet from the high side base of the tree) measured to a four inch top diameter inside the bark. Periodic improvements in the US Forest Survey concentrate on ever more statistically reliable measures of the stock, or "inventory," that meets these specifications despite evidence from the National Forest System that it is the economics, not the biology, of the specification that is weak. For example, in 1997 eight percent of all timber offered for sale was not bid upon. Surely this eight percent was uneconomic. Furthermore, the National Forest System lost $89 million on the timber that did sell-largely because the Forest Service absorbed the costs of building roads to the timber (USDA Forest Service 1998).[3] Apparently, the uneconomic share of the resource inventory is very large! Recognition of this uneconomic nature of many harvests has played a substantial role in environmental arguments to restrict Forest Service harvests, and in the Forest Service's own recent decision to reduce harvest levels.

We can demonstrate the problem with a simple diagram of the agricultural and forest landscape. Consider agricultural land first. The value of agricultural land is a function of the net farmgate price of agricultural products—which is greatest near the local market at point A in figure 6.1. Land value declines with decreasing access (which is closely related to increasing distance) as described by the function V_a. The function C_r describes the cost of establishing and maintaining secure rights to this land. This function increases as public infrastructure and effective control decline with decreasing access while the cost of excluding trespassers increases dramatically until no number of forest guards can fully exclude illegal loggers and other users of remote forests.

The functions explaining agricultural land value and the cost of secure property rights intersect at point B. Farmers manage land between points A and B for sustainable agricultural activities. They use land between points B and C (where agricultural land value declines to zero) as an unsustainable open access resource to be exploited for short-term advantage. They harvest native crops that grow naturally in this region, crops like fodder for their animals, native fruits, and fugitive resources like wildlife. They will not invest even in modest improvements in the region between B and C because the costs of protecting investments are greater than the investments are worth.

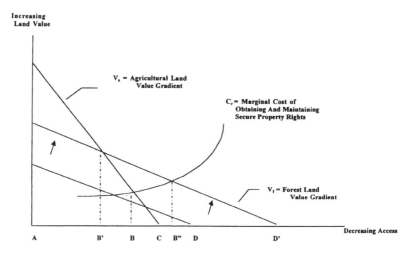

Figure 6.1. The Forest Landscape

The mature natural forest at the frontier of agricultural development at point B has an initial negative value because it gets in the way of agricultural production and its removal is costly. Later, when the forest first begins to take on positive value, it must be worth less than agricultural land and the function V_f describing forest value must intersect the horizontal axis at some distance beyond B. Market demand for forest products justifies their removal at this time, and it continues to justify their removal as the forest frontier eventually shifts out toward some point like D. At this point the market price of forest products just equals the cost of their removal and delivery to the market. The in situ price (the stumpage price of timber, for example) at this point is zero, and the value of forestland at D is also zero. The region of unsustainable open access activities now extends from B out to either C or D, whichever is farther. The costs of obtaining and protecting property rights insure that this region will remain an open access resource. Some governments protect some lands past point B but they must absorb the protection costs—and even then trespass occurs. For example, some amount of illegal logging occurs almost everywhere in the world. And some local citizens illegally harvest Christmas trees from the well-managed national forests near our residence in Virginia.

The construct of figure 6.1 conforms with the common description of any initial settlement. In some cases, trees actually impede agricultural development, the forest rent gradient is very low, and point D can even be to the left of point C. Apparently this describes Cote d'Ivoire (Lopez 1998) and some of Bolivia's Amazon region. The region between B and D can be large (Nepal's hills, India's Rajastan) but the forest in this region is generally degraded. Some trees in the region will re-grow naturally. The lowest wage households will continue to exploit these resources as the scattered trees grow to a minimum exploitable size (Amacher et al. 1993, Foster et al. 1998).

As the natural forest is depleted over time, the forest margin at D gradually extends farther and farther from the market and the delivered costs of forest products rise until they become sufficient to justify an alternative. The alternative may be non-forest consuming technologies (for example, kerosene or improved stoves), or it may be planting and sustainable management on some land closer to the market. The forest rent gradient rises with the increase in delivered costs (following the arrows in the figure) until it intersects the agriculture rent gradient to the left of the agriculture intersection with the cost function for property rights. Higher forest product prices now justify new planted and managed forests in the region B'B". These forests may take the form of industrial timber plantations or they may take the form of a few trees planted around individual households. The latter are excluded from most measures of the forest stock but their economic importance can be large. In Bangladesh, they account for 3/4 of all market timber and fuelwood consumption (Douglas 1982). Undoubtedly they account for an even larger share of all economic wood if we add subsistence collection for domestic consumption. In other examples, they are major sources of fuelwood consumption in Malawi (Hyde and Seve 1993), timber production in Kenya (Scherr 1995), and of positive environmental externalities in northern China (Yin and Newman 1996).

In all cases, commercial removal of the mature natural stock is concentrated around point D– or D' in regions characterized by the higher forest rent gradient. The harvest neighborhood of D (or D') can vary from year to year as investments in infrastructure (especially roads), forest policies, and relative prices shift the forest rent gradient at different annual rates. In most cases a mature natural forest of no economic value exists beyond D (or D'). Sometimes this region is small (e.g., in Ireland). Sometimes it continues well beyond point D (Siberia, Alaska, northern Canada) until it can become much the largest share of reported physical stocks.

The discussion to this point has taken a dynamic view, but consider what it means to measures of the forest at any moment in time—a static view.

Because most basic forest products are either bulky or perishable, their primary markets tend to be local. Therefore, in any country we observe many markets described by figure 6.1 and the actual physical stock of standing forest in most countries at any moment will include some from each of three categories. Reported stocks at any moment generally include a measure of plantations and household stocks (B'B"), another measure of degraded open access stocks (B or B" to D or D'), and yet another measure of mature natural stocks (D or D' and beyond).

Market and policy changes cause one time shifts in one or another of the three functions—V_a, V_f, and C_r and, therefore, one time shifts in the levels of one or another of the three stocks. Estimates of the impacts of these changes depend on how much of each of the three categories is included in the measured forest stock. Consider two identical countries, one that accurately measures all managed forests and includes smallholder tree crops (B'B") and one that only captures industrial plantations in its managed forests. Empirical assessments will (incorrectly) show that the forests of the former are more responsive to both forest management assistance and spillovers from agricultural policies. Consider another two identical countries, one that accurately measures all industrial plantations of perennial crops and one that measures plantations for wood fiber but excludes orchards and crops like rubber, palm oil, coconut, etc. The former captures a national perspective of the effects of any policy change. The latter more accurately captures the effects of policy change on the specific responsibilities of most forest ministries. Finally, consider a third set of identical countries, one that accounts for twice what the other accounts for in purely uneconomic physical stocks beyond the frontier at D (or D'). The country with the larger measure of stock will have a smaller rate of deforestation. These final two countries will receive differential attention by donors and the international environmental community entirely on basis of their different accounting for wholly uneconomic stocks. It is a matter of fact that the parts of our three forested regions that are included in official forest stocks vary from country to country, and even across regions within countries.

The remainder of this chapter is an empirical examination of this lesson. We will look for two fundamental improvements in the common physical measures of the forest stock: i) uniform and complete physical measures, incorporating all three stocks down to the same physical minimum and ii) modifications of these physical stocks to account for differences in economic market access. We will address these points with data from the forested regions surrounding 31 cities in eight countries in Africa, Asia, and Latin America. We will address the first point by measuring biomass, which is a more complete accounting of all fiber than common estimates of commercial-sized timber growing in commercial-sized blocks, and the second by incorporating distinctions in location, infrastructure, and terrain on the supply-side of our analytical model.

The Formal Model

Consumptive uses of the forest might be summarized as the collection of i) demands for wood fiber (timber, pulpwood, fuelwood, etc.) plus ii) demands for the conversion of forestland to agricultural uses. Consider both demands as demands for forestland. The demand for land to produce wood fiber is a function of the price for fiber delivered to the local market. Similarly, the demand for land to be converted to agriculture is a function of the delivered price of agricultural crops. The local market is the point of reference for prices regardless of whether the forest or agricultural products are consumed locally or shipped afar. Supply and demand must be measured at the same point. Therefore, supply for either use is the supply perceived by an entrepreneur contemplating its use from his/her position at the market center. This means that the in situ supply of standing timber must be adjusted for the costs of harvest and shipment to the local market.

More formally, demand for fiber Q_{df} is a function of the delivered price of fiber p_f, the price of other goods p_o, the per capita level of income I, and perhaps the regional population. Lower fiber prices and larger populations increase the quantity demanded. The sign on the income variable is less certain. The aggregate of various consumptive demands for wood fiber probably increases as per capita income rises to some level. It may decrease for incomes above that level.[4] It is also unclear whether substitutes or complements dominate the effects of other goods.

$$Q_{df} = Q_{df}(p_f, p_o, I, p) \qquad (6.1)$$

A variety of substitute energy and land use policies and local institutional arrangements also affect demand. We will return to these momentarily.

Agriculture is the greatest source of demand Q_{da} for forest clearance. Demand for agricultural land conversion increases with increasing agricultural prices and expanding populations. Once more, the effect of increasing per capita income is unclear. The net effect of substitutes and complements is also unclear.

$$Q_{da} = Q_{da}(p_a, p_o, I, p) \qquad (6.2)$$

The supply of standing forest for consumptive uses Q_{sf} is a function of delivered agricultural and forest product prices, the standing forest stock x, and its access to the market. Access explains the costs of delivery to the market. Access is affected by distance d, road density in the harvest region r, terrain, and moisture m. Moisture is an input to larger stocks but it is also an important deterrent to the use of some forestlands in some seasons. Higher prices, larger stocks, shorter distances, greater road density, more regular terrain, and increased moisture each increase the share of the standing physical stock available for use. The same arguments explain the supply of forestland for clearance and agricultural conversion Q_{sa}. (In particular, greater agricultural prices increase forest conversion and greater wood fiber prices increase fiber removal and make it easier to convert land from forests to agricultural uses.) Therefore, the sum of supply for both classes of forest use is

$$Q_{sf} + Q_{sa} = Q_s (p_f, p_a, x, d, r,, m) \qquad (6.3)$$

We can combine agricultural and wood fiber prices as a ratio such that it becomes changes in this ratio that affect the demand and supply of forests and forestland. The reduced form of our system of equations now becomes

$$P_f / p_a = p (p_o, I, p, x, d, r,, m) \qquad (6.4a)$$

$$Q = Q (p_o, I, p, x, d, r,, m) \qquad (6.4b)$$

We are interested in eq. (6.4b), the reduced form quantity equation which we can simplify further by treating the price of other goods as the numeraire and eliminating it from the estimation procedure.

Let's first return, however, to the policies and institutional arrangements we promised to revisit. Two classes of these may affect consumption opportunities from forest stocks at the margins of our three categories of forest, i) direct forest protection and ii) the broader class of other policies and institutions that affect the forest stock less directly. The first includes parks, wildlife preserves, and other preserved forest areas π_1. The second might include, for example, kerosene and fertilizer subsidies that alter fiber demands and agricultural production, respectively, and thereby modify the draws on the stock of forestland. It includes the institutions that control the establishment and transfer of property rights. And it includes macroeconomic policies that favor capital, disemploy labor, and thereby induce migration to undeveloped forested regions. In an analysis of improved measures of supply we are not interested in distinguishing between the many and complex policy and institutional determinants of deforestation and reforestation. Therefore, we will only identify them in the aggregate as π_2.

Our final formulation for the magnitude of the forest stock at any moment in time is

$$Q = Q\,(I,p,x,d,r,,M;\pi_1,\pi_2) \qquad (6.5)$$

Forest growth is not an issue because we are examining the stock at a moment in time. Forest growth would be relevant for questions of intertemporal adjustment, but questions of intertemporal adjustment would also have to incorporate rates of change for each of the other independent variables.

Data

Table 6.1 identifies the cities and countries in our sample. The intensity of effort required to map the forest cover and calculate the standing forest volumes around these cities limited our sample. Therefore, we selected cities and countries to reflect a range in both population and forest density. They also reflect regions in which colleagues at the Industry and Energy Department of the World Bank had personal experience. We used this experience to check our selection of maps and our calculations of standing forest volume, and to confirm our choices of other input data. The cities in our sample ranged in population from 20,000 to seven million. The forest stocks in the regions around them ranged from 768,000 cubic meters to 184 million cubic meters of woody biomass.

Table 6.1. Countries and Urban Centers in the Sample

Indonesia	Botswana	Bolivia
Bandung	Francistown	La Paz
Jakarta	Gabarone	Oruro
Semarang	Selebi-Phikwe	Quillacolo
Surabaya		Tarija
Yogyakarta	Mauritania	Trinidad
	Atar	
Philippines	Kaedi	Haiti
Bocolod	Kiffa	Port-au-Prince
Cagayan de Oro	Nouadhibou	
Cebu	Nouakchott	
Davao		
Manila	Zimbabwe	
	Bullawayo	
Thailand	Harare	
Ayutthaya	Masvingo	
Bangkok	Mutare	
Chiang Mai		

112   Economics of Forestry and Rural Development

Satellite maps of forest cover were the source of our raw data on standing forest stocks. We used digitized computer mappings from the GEMS study (FAO 1981) and the Atlas Draw/Atlas Graphics software package to calculate standing forest volumes for various categories of vegetation, and then converted these volume measures to estimates of total above-ground biomass, including the crown and understory as well as the bole of the tree. For our analysis, the important features of the calculation are the incorporation of all biomass and the uniformity of the calculation across all land areas in the sample.

We calculated biomass volumes for five concentric zones ranging up to 300 km around each urban center but we learned to disregard the latter two concentric zones, distances beyond 100 km. It seems that urban impacts decline substantially beyond 100 km and the overlapping effects of adjacent urban centers begin to confuse the analysis beyond this distance. Our estimates of standing biomass ranged from zero for the first zone of eight cities to over a million cubic meters for the third zone of nine cities. Table 6.2 shows the biomass volumes for the zones around one city, Cagayan de Oro in the Philippines, as an example.

The remaining independent variables are moisture, per capita income, population, road density, terrain, and the measures of policy and institutional differences. Moisture is average rainfall in mm/yr within a zone. Per capita income (in US $/yr) and population refer to the urban center. They are the same for all concentric zones associated with a city. Road density is the number of kilometers of primary road in a zone divided by the total area of the zone in ha. Terrain could be measured a number of ways (e.g., the difference in elevation extremes within a zone, the standard deviation in elevation difference among several sample points within a zone, or some threshold of elevation change within a zone). A dummy variable identifying the 350-meter threshold of elevation change performed best in our econometric analysis. Apparently, minimal change in topographic relief is no deterrent to forest harvest or land clearing, but some level of topographic variation in the neighborhood of 350 meters within a zone is a threshold that begins to constrain these activities. The Times Atlas of the World (1986) was our source for the moisture, population, road density, and terrain variables. The World Bank (1986) was our source for income data. Table 6.3 provides an impression of the variation in our data. Finally, a dummy variable for each country summarizes the collection of national policies and institutions that affect the level of the forest stock.

Results

Table 6.4 shows our OLS regression results based on a sample of 93 observations (31 cities, each with three zones). The independent variables are

Table 6.2. Biomass Table for Cagayan de Oro (in m³)

Zone	Distance (km) from urban center	Closed broadleaf forest	Open broadleaf forest	Pine forest	Mangrove	Plantations	Total woody biomass
1	0–25	1,561,623	0	0	0	551,039	2,558,565
2	26–50	29,460,715	6,745,148	0	0	608,050	34,450,989
3	51–100	107,016,560	72,123,727	0	0	2,033,338	183,597,155
4	101–200	147,437,124	235,981,464	0	0	7,669,456	403,851,683
5	201–300	90,901,754	130,033,288	0	490,076	8,441,170	237,466,377

the determinants of standing forest biomass (in m^3) in each concentric zone around the urban centers. The estimated equation itself is satisfactory. That is, a joint misspecification test cannot reject the null hypothesis that the sample data are normal and homoskedastic, the linear functional form is appropriate, and the coefficients are stable. The equation F test and adjusted R^2 are satisfactory. The coefficients on the basic demand and supply variables support our expectations and all five of these coefficients are statistically significant. Positive signs on these coefficients indicate that increases in the independent variables increase the volume of standing forest. Negative signs indicate that increases in the independent variables decrease this volume, or contribute to deforestation.

The unambiguous sign on the coefficient of per capita income is most interesting. We speculated that this sign could be either positive or negative, as the dominant opinion is that forest stocks first decrease, then increase, with increases in per capita income. Our significant positive sign suggests that forest stocks increase with per capita income even at the lower income levels characteristic of the countries in our sample.

The signs and significance for our supply-side variables are encouraging of our argument to correct the physical stock for other factors influencing supply. Moisture and rough terrain deter access, thereby leaving a larger standing physical stock. Roads enhance access, thereby leaving a smaller stock.[5]

The remaining independent variables refer to national policies and institutions. Forest preserves remove land area and forest volume from the stock available for exploitation in order to protect some feature of the forest environment. Therefore, successful forest preserves effectively concentrate forest exploitation on the forest area outside the preserve and reduce this available standing volume. Our measure of forest preserves has the expected sign, but the effect is not large and the sign is not significant. The probable explanation is that most preserves extend over large land areas and it is almost impossible to exclude some local forest consumers from them. Certainly the forest ministries and park agencies that manage the preserves of most developing countries have problems with trespass and theft. Some consumers extract biomass from forest preserves despite their protected status and some subsistence farmers even manage to convert large preserved areas to agriculture.

Country dummy variables capture the shift in the regression intercept due to differences from an arbitrary base case, Zimbabwe in our regression. This shift could reflect policy, institutions, infrastructure, or cultural differences. It is always negative in this regression-which only means the comparison with Zimbabwe is always negative. The most important infrastructure for forestry is probably roads. We captured the road effect

Table 6.3. Sample Characteristics (mean observations for urban centers in each country)

	Range of biomass (across zones and cities—in 000s m³)	Mean rainfall (mm/year)	Range in per capita income (US$)	Range in urban population (000s)	Range in road density (km/10⁶ ha)	Range of std. dev. elevation (meters)
Indonesia	0–7,152	1,689	250–462	507–7,202	243–1,068	0–762
Philippines	379–183,597	1,976	198–663	291–7,567	24–1,249	0–652
Thailand	0–194,805	1,320	630–1,061	37–4,329	411–966	0–280
Botswana	3,326–61,607	466	596–1,342	35–96	161–510	0–257
Mauritania	0–17,637	227	176–476	0	0–715	0–104
Zimbabwe	37,644–123,508	769	404–882	41–785	207–858	0–289
Bolivia	0–301,390	545	940–1,726	23–677	15–625	0–1,104
Haiti	768–11,479	1,500	949	513	457–720	273–528

116   Economics of Forestry and Rural Development

Table 6.4. Determinants of Standing Biomass—Within Concentric Zones around an Urban Center

Variables	Coefficients	Country dummies, p_g	Coefficients
Constant	111.19* (3.82)	Philippines	-16.99 (-0.91)
Moisture, m	0.017*** (1.85)	Bolivia	-25.90*** (-1.85)
Per capita income, I	0.026*** (1.80)	Indonesia	-35.85** (-2.18)
Population, p	-12.30** (-2.28)	Botswana	-41.88* (-2.89)
Road density, r	-0.047* (-3.37)	Thailand	-42.79* (-3.45)
Terrain, _	33.06* (4.16)	Mauritania	-51.22* (-3.78)
Forest preserves, p_1	-9.86 (-1.39)	Haiti	-80.09* (-4.08)
adjusted R^2 F (13, 82)			0.586 11.34*

*, **, ***Significance at the 0.01, 0.05, and 0.10 level, respectively. Numbers in parentheses are asymptotic t ratios.

explicitly. Therefore, our coefficients mean that policy, institutional, or cultural differences caused the other seven countries to decrease their forest stocks below levels characteristic of Zimbabwe. For some countries we can speculate on the source of this difference. In Thailand, it may have been the military's active encouragement of migration into the forested northeast. For Indonesia, it may have been the transmigration resettlement policy. For other countries, it may not have been a forest-related policy, institutional, or cultural activity. It may have been the overall macroeconomic environment. Certainly an unstable and declining economic environment contributed to Haiti's extreme level of deforestation.[6]

Conclusions

Our contention has been that the difference between physical and economic measures of resource stocks is critical for policy analysis. This point is fundamental and well-known. The differences in physical measures between countries and between classes of a resource even within a country are less commonly recognized. And estimates of the magnitude of error in policy analysis due to overlooking these differences are unknown to us.

The public agencies that collect the physical data have their own responsibilities and data collection is expensive. Therefore, we can anticipate their greater emphasis on data that address their own responsibilities. For example, forest ministries will concentrate on measures of mature natural stocks beyond the frontier. The lands at the frontier and beyond are their responsibility. Private lands and private stocks, especially many perennial tree crops and agroforestry fall outside the responsibility of most forest ministries.

Most resource management agencies are also interested in continuous measures, in forestry "continuous forest inventories." These are measures whose base units do not change over time. Continuous measures are necessary for tracing intertemporal patterns of use of the natural stocks but they are at odds with economic evaluations because market and policy effects change even as measures of physical stocks do not.

Policy analysts are reliant on these agencies and their data because it is generally too expensive to conduct an assessment like our GIS assessment of biomass as part of each new analysis. Policy analysts can make our second improvement, however, by introducing the economic characteristics of supply appropriate to their analyses. The R^2 statistic in our regression illustrates the importance of this improvement. This statistic identifies the share of all variation in the dependent variable that is explained by the independent variables. Our R^2 was approximately 0.60. The same regression without the road, terrain, and moisture variables produced an R^2 of 0.20—and both regressions already reflect the correction for distance in their dependent variable. Therefore, factors that characterize the difference between physical stocks and economic supply explain at least forty percent of all variation in the resource stock.

This is strong evidence—but we can develop the argument even further. We can contrast this "forty-percent" finding with an estimate of the importance of policy and institutional factors on the resource stock. The country dummy variables capture these factors, albeit in an aggregate manner. They explain only twenty percent of the variation in resource stocks.[7] Evidence from an altogether different source supports an argument that twenty percent is in the right range for the effect of policies and institutional differences on the stocks. Deacon (1994) examined the links to deforestation in a cross-section of 120 countries. He measured resource flows not stocks, and he was reliant on physical inventories collected by national governments. He found that policies and institutions explain approximately twenty percent of the variation in his resource flows. If a twenty percent variation in the flows, over time, adds up to a twenty percent variation in stocks, then Deacon's evidence and ours are remarkably consistent.

The forty percent of our variation explained by economic factors affecting the resource supply dominates the twenty percent explained by policies and institutions. This is a critical finding for those who design resource inventories, for policy analysts, and for decision-makers alike. It means that measurement errors generally dominate policy effects and it means that even modest economic adjustments in measures of the stock will improve policy expectations. Our findings on this point are so compelling and so important that, to us, they beg further inquiry—elsewhere in forestry and for other natural resource commodities as well. Are such differences between measures of economic and physical stocks the norm? And how damaging are these differences for policy analysis? Our evidence also urges extreme caution on acceptance of policy analyses that fail to make economic corrections in physical stocks. For forestry, these include most non-GIS-based assessments of deforestation rates and policy impacts on deforestation, many assessments of carbon sequestration and global change, and most empirical evidence either supporting or rejecting environmental Kuznets' curves.

NOTES

1. School of Public and Environmental Affairs, Indiana University; Chinese Academy of Agricultural Science; Industry and Energy Department, World Bank; and Centre for International Forestry Research; respectively. Marnie Taylor calculated our measures of biomass. R. Deacon and an unidentified reviewer shared critical insights on earlier drafts of this paper. An earlier version of this paper appeared as "The importance of economic corrections for physical stocks: a forestry example" in *Forest Ecology and Management*.

2. In a recent review of the frontiers of environmental and resource economics, Deacon et al. (1998) expressed the opinion that "The spatial dimension of resource use may turn out to be as important as the exhaustively studied temporal dimension in many contexts." They went on to emphasize the importance of resource boundaries and they gave examples from non-market uses of the forest. Later in a paper on forest economics research in the same review volume, Parks et al. (1998) focused on the effects of location and cited some of the recent literature that has begun to respond to our problem. Chomitz (Chomitz and Gray 1996) has been particularly attentive to issues of land quality and road access with regard impacts on development and deforestation.

3. Hyde (1980) discusses the specifics of the US Forest Service calculation in greater detail.

4. The general expectations are that the income elasticity for the paper production demand on wood fiber is greater than one, the income elasticity for the construction demand on wood fiber is positive but less than one, and the income

Estimates of Economic Supply from Physical Measures of the Forest Stock 119

elasticity for fuelwood is close to zero even in subsistence economies. The income elasticity of non-consumptive aesthetic demands on standing forests is greater than one. This final effect may eventually offset the first three and cause standing forests to increase as incomes rise beyond some level.

5. We had no reason other than statistical fit to choose between absolute values and natural logs in our estimation. Others (e.g., Deacon 1994) have argued that natural logs of population and income are preferred when comparing countries with a greater range in per capita incomes.

6. Deacon compared the coefficients on our county dummies with average investment per GDP and average investment per worker for the same countries in the period 1976–85. His correlation coefficients of .3078 and .4753, respectively, lend conviction to our ordered ranking of countries and to the hypothesis that macroeconomic differences explain some level of investment or disinvestment in forestry.

7. There may be an endogeneity problem as the level of the stock may affect policy as well as policy effects the level of stocks. The endogeneity problem should be small in comparison with the mismeasurement problems we have discussed, however, in the common case where the physical stocks far exceed the economic stock.

REFERENCES

Alston, L., G. Libecap, and R. Schneider. 1996. "The Determinants and Impact of Property Rights: Land Titles on the Brazilian Frontier." *Journal of Law, Economics, and Organization.* 12(1): 22–65.

Amacher, G. S., W. F. Hyde, and B. R. Joshee. 1993. "Joint Production in Traditional Households." *Journal of Development Studies* 309(1): 206–225.

Bowles, I. A, E. R. Richard, R. A. Mittermeier, and G. A. B. Fonseca. 1998. "Logging and Tropical Forest Conservation." *Science* 280 (19 June): 1899–1900.

Brown, S., and A. Lugo. 1984. "Biomass and Tropical Forests: A New Estimate Based on Forest Volumes." *Science* 223 (23 March): 1290–93.

Bryant, D., D. Nelson, and L. Tangley. 1997. "The Last Frontier Forests: Ecosystems and Economies on the Edge." Washington: World Resources Institute.

Chomitz, K. M., and D. A. Gray. 1996. "Roads, Land Use, and Deforestation: A Spatial Model Applied to Belize." *World Bank Economic Review* 10(3): 487–512.

Deacon, R. T. 1994. "Deforestation and the Rule of Law in a Cross Section of Countries." *Land Economics* 70(4): 414–430.

Deacon, R. T. 1995 "Assessing the Relationship between Government Policy and Deforestation." *Journal of Environmental Economics and Management* 28(1): 1–18.

Deacon, R. T., D. S. Brookshire, A. C. Fisher, A. V. Kneese, C. D. Kolstad, D. Scrogin, V. K. Smith, M. Ward, and J. Wilen. 1998. "Research Trends and Opportunities in Environmental and Natural Resource Economics." *Environmental and Resource Economics* 11(3-4): 383–397.

Douglas, J. J. 1982. "Consumption and Supply of Wood and Bamboo in Bangladesh." Field document no. 2, UNDP/FAO project BGD/ 78/010, Bangladesh Planning Commission, Dhaka, Bangladesh.

FAO/UN (Food and Agricultural Organization of the United Nations). 1981. "Tropical Forest Resources Assessment Project (in the framework of GEMS)," Forest Resources of Tropical Asia. Rome: FAO.

Foster, A. D., M. R. Rosenzweig, and J. R. Behrman. 1977. "Population Growth, Income Growth, and Deforestation." Draft ms., Economics Department, University of Pennsylvania.

Howe, C. W., 1979. *Natural Resource Economics* New York: John Wiley.

Hyde, W. F. 1981. *Timber Supply, Land Allocation, and Economic Efficiency Baltimore*: Johns Hopkins University Press for Resources for the Future.

Hyde, W. F., and J. Seve. 1993. The Economic Role of Wood Products in Tropical Deforestation: the Severe Example of Malawi." *Forest Ecology and Management* 57(2): 283–300.

Krutilla, K., W. F. Hyde and D. F. Barnes. "Periurban Deforestation in Developing Countries." *Forest Ecology and Management* 74(2): 181–95.

Lopez, R. 1998. "The Tragedy of the Commons in Cote d'Ivoire Agriculture." *World Bank Economic Review* 12(1): 105–132.

Openshaw, K. 1986. "Methods of Collecting Biomass Supply Statistics." Washington: World Bank working paper.

Ostrom, E. 1995. "A Framework Relating Human 'Driving Forces' and their Impact on Biodiversity." Bloomington, IN: Draft manuscript, Workshop in Political Theory and Policy Analysis, Indiana University.

Parks, P. J., E. B. Barbier, and J. C. Burgess. 1998. "The Economics of Forest Land Use in Temperate and Tropical Areas." *Environmental and Resource Economics* 11(3-4): 473–487.

Powell, D. S., J. L. Faulkner, D. Darr, Z. Zhu, and D. MacCleary. 1993. *Forest Resources of the US, 1992*. Washington: USDA Forest Service General Technical Report RM 234.

Scherr, S. J. 1995. "Economic Factors in Farmer Adoption of Agroforestry: Patterns Observed in Western Kenya." *World Development* 23(5): 787–804.

Times Atlas of the World. 1986. Edinburgh: Times of London in collaboration with John Bartholomew & Son.

von Thunen, J. H. 1875. *The Isolated State in Relation to Land Use and National Economy*. Berlin: Schumaucher Zarchlin.

USDA Forest Service. 1998. *Forest Management Program Annual Report*. Washington: USGPO.

World Bank. 1986. *World Development Report*. Washington: World Bank.

Yin, R. and D. H. Newman. 1996. "The Impacts of Rural Reforms: The Case of China's Forestry Sector." *Environment and Development Economics* 2(3): 289–304.

CHAPTER 7

Regional Fuelwood Production and Consumption in Nepal: With Implications for Local Adoption of New Forestry Practices

*Keshav R. Kanel, Gregory S. Amacher,
William F. Hyde[1], and Lire Ersado*

Fuelwood is a fundamental source of rural energy—and of deforestation—around the world. Rural households often cannot afford commercial energy substitutes like kerosene, even where they are available. They may use agricultural residues and improved stoves, a technological substitute for fuel. Residues, however, are economically inferior substitutes and they have important alternative roles in maintaining soil quality and as livestock food supplements; and improved stoves have had a mixed history in fuelwood substitution programs. Many rural households remain dependant on fuelwood for heating and cooking.

Development agencies have been interested in the roles of fuelwood in rural energy and deforestation since the 1970s, but the empirical economics literature on fuelwood is more recent and largely restricted to household evidence from selected local areas of Nepal (Bluffstone 1995, Cooke 1998, Amacher et al. 1993). Our objective in this chapter is to develop broad evidence from Nepal's two major inhabited regions and to examine local variation across 27 districts within the two regions. We will develop information on the regional characteristics most suggestive of relative fuelwood scarcity, and on the characteristics of those households and districts that suffer most severely from it. In general we will find that the fuelwood scarcity problem is not as universal as often perceived—but scarcity is an important determinant of behavior for some households in some districts.

We will begin with a model of household behavior and then derive the household supplies of fuelwood for both internal use and market sale, as well as the household's market demand. Market production is a potential income source for the poorest households. It has not been examined previously, although Bluffstone (1995) suggested that an exogenous labor market (and, by implication, a market for the services of fuelwood collectors) can eventually support levels of plantation fuelwood production and market energy substitution that will offset a history of increasing deforestation.

Our regional evidence suggests that market prices are important but insufficient indicators of fuelwood scarcity wherever consumption includes households that do not participate in the market. In general, those fuelwood collecting households that do not participate in the market also do not respond to local economic conditions in the same way that market participants respond. Therefore, successful forest policy must combine local market information with evidence on household behavior in order to identify the districts where fuelwood is, indeed, scarcer—and to suggest successful responses to deforestation. We will find that consumption interventions like improved stoves or production interventions like the distribution of new seedlings are more likely to be successful where the substitution elasticities in consumption or production, respectively, are greater. Both are more likely when targeting the most price responsive fuelwood collecting households. In Nepal, these households are situated in select districts of the hill region—rather than in the broad tarai region where forest stocks actually tend to be smaller.

A Model of Household Fuelwood Behavior

The production and consumption decisions of rural households are often interdependent because household labor is a key input for agricultural production and the returns from the household's agricultural activities constitute an important share of the wealth that households have available for consumption purposes. Households hire outside labor to assist in their agricultural activities (Amacher et al. 1993), but households in our sample seldom use hired laborers for fuelwood collection. Therefore, hired labor is not a perfect substitute for household labor in production, and the price of household labor services is an endogenous virtual price that is unknown except to each household itself. This means that specifications of the household production and consumption decisions are non-separable (Singh et al. 1986).

More formally, representative households maximize a continuous, monotonic, quasi-concave utility function subject to budget, time, and non-negativity constraints. Household utility U is a function of the goods and services Q that use fuelwood F and agricultural residues R as inputs, household labor L, the consumption of other goods X, and various local demographic characteristics Ω that are important to household preferences.

$$\max_{Q,L,X} U(Q, T-L, X; \Omega)$$
$$\text{s.t. } p_x X + p_f F_p + p_r R = p_f F_s + I$$
$$Q \geq 0 \quad (7.1)$$
$$F_c \geq 0, F_p \geq 0, F_s \geq 0; R \geq 0$$
$$L \geq 0, T-L \geq 0$$

where F_p, F_c, and F_s are fuelwood purchased, collected, and sold, respectively. T is total time and L specifically refers to labor allocated to fuelwood collection. The p_i are prices and I is exogenous income from all non-fuelwood sources, both on- and off-farm.

Fuelwood and residues are intermediate inputs in household utility

$$Q = Q(F_h, R; \Xi) \qquad (7.2)$$

where F_h is total fuelwood consumption in the household, and Ξ is a vector of technologies that affect the efficiency of fuelwood and residue consumption (e.g., stove quality). The complete set of strategies with respect to fuelwood allows households to collect, purchase, or selling. Total household consumption is

$$F_h = F_c + F_p - F_s \qquad (7.3)$$

The fuelwood production function F_c is continuous and quasi-concave in its arguments:

$$F_c = F_c(L, A; \Omega) \qquad (7.4)$$

where L and A are vectors of the variable and fixed factors of production, respectively. Fuelwood collection is a labor intensive activity.

Substituting eqs. (7.3) and (7.4) into the budget constraint yields the production augmented budget constraint.

$$p_x X + p_f F_p + p_r R = p_f F_s(L, A; \Omega) + I \qquad (7.5)$$

The household maximizes utility subject to time, non-negativity, and the new budget constraint. The first order conditions for utility maximization are

$$\begin{aligned} &\partial U/\partial X = \lambda p_x \\ &\partial U/\partial L = \lambda p_x\, \partial F_s/\partial L + \mu \\ &[\partial U/\partial Q][\partial Q/\partial F_p] = \lambda p_f \\ &[\partial U/\partial Q][\partial Q/\partial F_s] = -\lambda p_f \\ &[\partial U/\partial Q][\partial Q/\partial R] = \lambda p_r \\ &\lambda p_f\, \partial F_c/\partial L = 0 \end{aligned} \qquad (7.6)$$

and the budget constraint. λ is the Lagrangian multiplier associated with the marginal utility of income and μ is the Lagrangian associated with the constraint on household time. The second condition of eq. (7.6) shows that the price of household labor is an endogenous value equal to the value of the household's marginal product for labor used in fuelwood collection. The third and fourth conditions show that those households that purchase and sell fuelwood face a market price. The second-fourth conditions also show that those households that collect fuelwood for their own consumption face more complex decisions involving labor opportunities that depend on household preferences, technologies, and marginal utilities. Goetz (1992) argued and Amacher et al. (1996) showed that the opportunity costs for household fuelwood collection can be significantly different from the observed market fuelwood price. This means that conclusions about fuelwood scarcity that are solely reliant on market information may diverge from conclusions that also incorporate evidence on the household's collection for its own internal use.

The first order conditions and the budget constraint provide most of the information necessary to derive the household's labor supply for fuelwood production, its market purchase and supply of fuelwood, and the household's consumption of agricultural residues and other goods.

The final condition of eq. (7.6) implies that the household maximizes its net income conditional to its chosen level of labor. Following Thornton (1994), the household's net income (in cash terms) can be defined as

$$N(p_f, p_r, A;L) = G(p_f, p_r, A;L) - p_a Aa \qquad (7.7)$$

where

$$G(.) = \max \{p_f F_s(L, A;\Omega)\}$$

and p_a is a vector of prices of fixed inputs. $G(.)$ has the properties of a variable profit function. The revised household budget constraint becomes

$$p_x X + p_f F_p + p_r R = N(p_f, p_r, A;L) + I \qquad (7.8)$$

A set of output supply and input demand functions can be created by applying Hotelling's Lemma to eq. (7.7)

$$HF_h = \partial N/\partial P_f = HF_h(p_f, p_r, A;L) \qquad (7.9a)$$

$$HL = \partial N/\partial \omega = HL(p_f, p_r, A;L) \qquad (7.9b)$$

where HF_h and HL are the conditional net income maximizing supply and demand choices, and ω is the price of variable inputs (e.g., labor).

The consumption equations can be derived in the same manner. The revised budget constraint, eq. (7.8), leads to necessary conditions identical to (7.6), except that the second necessary condition takes the form

$$\partial U/\partial L = \lambda\, \partial N/\partial L$$

where $\omega = \lambda \partial N/\partial L$ is the "virtual" or shadow price of labor used in fuelwood collection since it is unobserved. Therefore, when the household maximizes its utility, $\partial N/\partial L$ is the value of the household's marginal product.

It is now clear that the utility maximization problem creates a set of household choices of the form

$$\begin{aligned} X &= X(p_x,\omega,p_f,p_r,I) \\ L &= L(p_x,\omega,p_f,p_r,I) \\ F_p &= F_p(p_x,\omega,p_f,p_r,I) \\ F_s &= F_s(p_x,\omega,p_f,p_r,I) \\ R &= R(p_x,\omega,p_f,p_r,I) \end{aligned} \qquad (7.10)$$

where the terms in ω contain λ.

The production and consumption sides of the model can be integrated by substituting the second condition of eq. (7.10) into the conditional net income maximizing supply and demand choices, eqs. (7.9).

$$HF_h = HF_h[p_f,p_r, A, L(p_x,\omega,p_f,p_r,I)] \qquad (7.11a)$$

$$HL = HL[p_f,p_r, A, L(p_x,\omega,p_f,p_r,I)] \qquad (7.11b)$$

Eqs. (7.10) and (7.11) are the structural equations of the model. It is clear from them that the household production and consumption decisions are interdependent. That is, changes in the exogenous factors of consumption induce changes in labor supply and, consequently, in the household's choices of inputs and outputs. Similarly, production shocks induce changes in the virtual price of labor, which affects the household's labor supply and its choices of consumer goods.

Empirical Specification

The empirical fuelwood production function corresponding to eq. (7.4) is

$$F_c = F_c(L,A,\Lambda,\Omega,\epsilon_c) \qquad (7.12)$$

where L remains the household's labor allocation for fuelwood collection, A is a measure of the household's fixed assets (including the standing community forest available to the household), and Λ is a measure of the standing forest resource available to the household, and Ω remains a vector of demographic variables. ϵ_c is a mean zero error term. We assume that eq. (7.12) is of Cobb-Douglas form.

The conditional profit function based on Cobb-Douglas production is linear in log-linear, as are the associated conditional output supply and input demand functions corresponding to eqs. (7.11)

$$HF_h = HF_h(\omega, p_f, p_r, A, \Omega, \epsilon_{hf}) \quad (7.13a)$$

$$HL = HL(\omega, p_f, p_r, A, \Omega, \epsilon_{hl}) \quad (7.13b)$$

We will follow Thornton (1994) in using the Stone-Geary form for our utility function and assess the expenditure form of eqs. (7.10). The Stone-Geary form satisfies the classic propositions of consumer theory (positive and declining marginal utilities for all goods). Maximizing this utility function subject to the non-linear budget constraint, eq. (7.8), creates a non-linear estimation problem that [following Thornton (1994)] can be solved by assuming budget linearity at the point of utility maximization. The predicted value of the household's marginal product of labor that emerges from this estimation procedure becomes a proxy for the household's unobserved virtual wage.

The empirical forms of the household choices corresponding to eqs. (7.10) now become

$$wL = L(\omega, p_f, p_r, A, I, \Omega, \epsilon_l) \quad (7.14a)$$

$$p_f F_p = F_p(\omega, p_f, p_r, A, \Omega, \epsilon_p) \quad (7.14b)$$

$$F_s = F_s(\omega, p_f, p_r, A, I, \Omega, \epsilon_s) \quad (7.14c)$$

$$R = R(\omega, p_f, p_r, A, I, \Omega, \epsilon_r) \quad (7.14d)$$

The ϵ_j are mean zero errors. We can make p_x the numeraire and remove it from the system. The virtual wage term in all conditions of eq. (7.14) can be calculated as $\omega = \gamma \, \hat{F}_c / L$, where γ is the estimated coefficient on the labor term in the production function and $\hat{F}_c$ is the predicted measure of household production. The labor demand and supply equations, (7.13b) and (7.14a), are less important to the fuelwood focus of our analysis and we will not develop their empirical estimates here.

Eq. (7.14b), the market expenditure function for fuelwood, is linear—as discussed. Eq. (7.14c) explains the household supply of fuelwood to the market. It is of log-log form since it was derived from the Cobb-Douglas production function. Eq. (7.14d) explains household consumption of combustible agricultural residues. Agricultural residues are a fuelwood substitute but they are not traded in local markets in either the hills or the tarai. The absence of residue markets means that we have no observations on residue prices. Therefore, we are restricted to a residue consumption function, rather than an expenditure function, for eq. (7.14d). We will estimate its linear form. The residue consumption function provides our only indication of the importance of substitutes between fuelwood and residues. In sum, eqs. (7.12), (7.13a), and (7.14b-d) are the family of equations for empirical estimation. Their stochastic forms assume additive disturbance terms with expected values equal to zero and finite variances. The structural links between the production and consumption sides of the household model suggest a 2SLS estimation procedure.

Some households only collect fuelwood, and do not participate in the market. This creates a censored data dependent variable that can be accommodated with a Tobit model for the fuelwood purchase and sale equations, eqs. (7.14b,c). Finally, we will correct all equations for heteroskedasticity that might be attributed to family size or other household behavioral differences.

Data

Our data originate with two surveys conducted by students under our guidance at Nepal's Institute of Forestry in Pokhara. Both surveys feature household fuelwood production and consumption for a one month period. The tarai survey consists of a random sample of households interviewed during the Desain holiday (October) in 1988. It includes 286 rural household observations distributed across twelve of the twenty geo-political districts in the tarai. The mid-hill survey was also taken during Desain, but in 1987. It includes 240 observations from fifty rural panchayats (villages) in fifteen of 39 mid-hill districts. Both surveys are large and their geographic distributions are broad.

The regional distinctions between the tarai and the mid-hills are important. The mid-hill region is Nepal's traditional population center. It is also the focus of world concern for Nepal's rapid deforestation. The tarai is the lower, warmer, and drier Gangetic plain. It was an underpopulated jungle until malaria control in the 1960s permitted human immigration, deforestation, and agricultural land conversion. The tarai still contains some forest, particularly in its western reaches, and it remains a source of fuelwood for the urban centers of the hills (Sheikh 1989). The tarai's forest resource is dwindling, however, as this region becomes Nepal's best source of agricultural production.

Climate is one source of regional differences in the forest resource stock, in agricultural opportunity, and in aggregate household fuel consumption. The larger landholdings of tarai households may be an additional source of wood fuel, as woody plants often grow along the edges of fields and terraces, and farmers have begun to plant trees in response to deforestation and increasing scarcity (Kanel 1996). Demographic differences between the largely immigrant tarai population and the historically stable mid-hill population may also explain behavioral differences in fuelwood production and consumption.

Table 7.1 reviews descriptive statistics for rural households and districts in the two regions. The forest stocks in this table do not include trees on private farm lands. Therefore, we might think of the stocks in this table as public forests. These public stocks are larger on per hectare and per capita bases in the mid-hills than in the tarai. Hill households also consume more fuelwood, and collect more combustible agricultural residues. Fuelwood collection times are shorter and fuelwood prices are somewhat smaller in the hills than in the tarai. These observations are consistent with the smaller forest inventories (per capita and per hectare) in the tarai and with Kanel's observations of more private reforestation activity in the tarai.

Table 7.1. Descriptive Statistics[a]

Statistics	Mid-hills	Tarai
Average district population	219,830	315,210
Land area per district, thousand ha.	175,173	183,587
Forest area per district, thousand ha.	70,034	78,600
Forest inventory per district, thousand m^3	5,564	5,487
Fuelwood price, rupees/kg.	0.47	0.59
Fuelwood collection time, hours per kg.	0.03	0.15
Fuelwood consumption per household, kg.	365.20	72.90
Residue use per household, kg.	277.88	7.75
Land area per household, ha.	0.93	3.27
Average family size	7.69	8.85

[a]The first four measures are from the 1991 Census of Nepal or the forest survey office of the Ministry of Forests and Soil Conservation. The remaining measures are from our surveys. All tarai statistics are signficiantly different from their mid-hill counterparts at least at the .10 level.

Table 7.2 defines the endogenous and exogenous variables used in our regressions. The endogenous variables are fuelwood collection time (a measure of household labor allocated to fuelwood collection), the virtual wage of labor, the market price for fuelwood, total fuelwood collected, fuelwood purchased in the market, fuelwood sold in the market, and residues consumed.

The exogenous variables are household capital, forest resource availability, household income and various demographic indicators of local behavior. Livestock is our measure of household capital. It is a reasonable proxy for all household capital because livestock, and especially draft animals, are the household's most valuable capital and households with more draft animals also tend to have more of all other forms of capital. The effect of the livestock variable in our equations is less certain, however, because some livestock food requires heating and, thereby, consumes fuel. Therefore, livestock are both productive capital and consumers of fuel.

Our data include both physical measures of the resource stock (volume and land area in forest and scrub) and two indicators of its access to collectors and final users. Both mature forests, and degraded forests and small woody shrubs, support fuelwood extraction. In addition, we previously observed that some households obtain wood from their own lands. Access to the resource or to the market may be measured by distance to the nearest road (tarai) or trekking trail (mid-hills), or distance to the nearest market where wood can be sold or purchased. Greater distance means less market access and probably means reduced human access to the local forest stock; thus, less pressure on the resource. Therefore, greater distance probably means more household collection, less market expenditure, and less market supply. Our data set contains no information on exogenous income, but landholdings should be a good proxy for the household's farm income. Wealthier households probably collect less of their own fuelwood and rely more on market purchases.

In addition, family size, raksi production, ethnic group, and improved stove ownership are household demographic variables. Larger families produce and also consume more fuelwood. Raksi is the local alcoholic beverage. It is heat-brewed by the Newer ethnic group, but consumed by all classes of society. Brahmins are the highest caste in this society of many castes plus several ethnic groups. Higher caste and larger families both generally correlate with greater wealth. Finally, improved stoves are a technological substitute for fuel.

Regional Household Results

Table 7.3 presents our Cobb-Douglas production estimates, eq. (7.12). These are the source of observations for our estimated virtual wage term in all subsequent equations.

Table 7.2. Variables

Variable	Definition
Exogenous Variables	
Landholding	Land area under household management, a proxy for housheold income and a potential source for fuelwood and residues
Forest area	Forest land area in the district, in 1000s of hectares
Shrub area	Scrub forest land area in the district, in 1000s of hectares
Forest volume	Standing forest volume in the district, in 1000s of m^3
Distance to trail	Distance in km. From household to the nearest trekking trail (in the mid-hills)
Distance to road	Distance in km. From household to the nearest road (in the tarai)
Distance to town	Distance in km. From household to the nearest village
Family size	Number of people living in immediate household
Animal units	Number of livestock owned by a household
Raksi	Indicator of Newar households. Newars are an ethnic group that uses fuelwood to brew raksi, the local alcoholic beverage.
Brahmin	Indicator of higher (Brahmin) ethnic class
District population	District population
Improved stoves	An indicator of household possession of improved stoves, a technological substitute for fuel
Endogenous Variables	
Collection time	Total collection time for fuelwood in hours per month
Wage	The instrumental variable estimate of the shadow wage
Fuelwood price	Fuelwood price in rupees per kg.
Fuelwood collected	Fuelwood collected by the household in one month in kg.
Fuelwood purchased	Fuelwood purchased by the household in the market in one month in kg.
Fuelwood sold	Fuelwood sold in the market by the household in one month in kg.
Residues consumed	Residues consumed by the household in one month in kg.

Table 7.3. Fuelwood Production Functions[a]

Explanatory variables		Mid-hills	Tarai
Constant		3.638***	-2.357
		(3.507)	(-1.330)
Total collection time, a measure of the labor input	(+)	0.999*	0.771***
		(1.669)	(2.570)
Animal units, a measure of household capital	(+)	0.149	
		(0.624)	
Forest volume	(+)	-0.063	0.223*
		(-0.356)	(1.641)
Shrub area	(+)	0.026	
		(0.149)	
Distance to nearest trail	(-)	-0.143	
		(-1.373)	
Distance to nearest town	(-)		-0.032
			(-0.639)
Landholdings	(?)	0.103	-0.010
		(0.964)	(-0.140)
Family size	(+)	-0.0046	0.022
		(-0.172)	(1.154)
District population	(?)		0.065
			(0.734)
log likelihood function		-403	-444
degrees of freedom		231	248

[a]Parenthetical expressions are expected signs and t statistics. ***, **, and * indicate statistical significance at the 0.01, 0.05, and 0.10 levels. These are log-log 2SLS regressions.

Constant returns are a possibility for both equations as we cannot rule out the null hypothesis that the sum of coefficient estimates is equal to one. The log likelihood functions indicate satisfying equations and eight of ten coefficients for which we can establish expectations have the anticipated signs. The important labor terms (total collection time) are positive and significant in both equations. The significance of the labor variables, in contrast with the insignificance of most other independent variables, only stresses the role of household labor in fuelwood collection and the lesser roles of all other inputs. Apparently, even the level of the natural forest stock has not yet become a critical factor in production across the breadth of the mid-hill region—although it is an important explanatory variable in the tarai. The emphasis on labor as the factor of production is consistent with general observations that fuelwood harvests and fuelwood shipment are manual activities—in Nepal and most other developing countries. It is also consistent with prior econometric evidence from more geographically selective surveys from Nepal's hills (Amacher et al. 1993, Cooke 1998).

The larger coefficient collection time for the hills indicates a greater marginal product of labor for the collection activity in that region than in the tarai. It is consistent with the greater consumption, smaller average collection time, and larger resource stocks that we observed for the hills (table 7.1).

Total Fuelwood Collection

Table 7.4 shows our regression estimates for the household's total fuelwood collection, eq. 7.13a, for the combination of household use and supply to the market. The log likelihood functions for these estimates indicate satisfying equations and thirteen of fifteen coefficients for which we can establish expectations have the anticipated signs. Because these are log-log equations, the estimated coefficients are also elasticities.

For the hills, the critical price and estimated wage terms have the expected signs and both are significant. The more elastic (and more significant) price and wage terms for the hills are consistent with the greater level of fuelwood consumption in that region and, presumably, the larger role of fuelwood in the budgets of hill households. (That is, households are generally more responsive to price changes in items that consume a larger share of the household budget—and the level of fuelwood consumption is greater while average income is smaller in hill households.) The level of collection in the hills is responsive to the level and the accessibility of resource stocks, as expected, and improved stoves are a technical substitute for fuelwood, also as expected.

All tarai coefficients have the anticipated signs but only the coefficient on private landholdings is significant. The relatively smaller and insignificant coefficients for the tarai price and wage terms are consistent with the reduced level of consumption and, therefore, the relatively smaller role of fuelwood in tarai household budgets. The positive and significant sign on landholding is consistent with our observation of smaller public forest stocks in the tarai and the suggestion that some tarai households have begun to grow trees on their own lands. The general observation of inelastic wages and prices and substitution opportunities in the tarai is consistent with the conclusion that fuelwood in the tarai is a household necessity and one with few substitutes, but one that, fortunately, absorbs only a small share of tarai household budgets.

Table 7.4. Total Fuelwood Collected[a]

Explanatory variables		Mid-hills	Tarai
Constant		2.961** (2.433)	-2.484 (-1.130)
Estimated wage	(+)	1.329*** (3.873)	0.088 (0.087)
Fuelwood price	(+)	1.569** (2.052)	0.705 (1.304)
Forest area	(+)	0.446*** (2.722)	0.260 (1.321)
Shrub area	(+)	0.205 (0.999)	
Distance to nearest trail	(+)	0.221** (2.500)	
Distance to nearest town	(+)		-0.057 (-0.085)
Landholdings	(?)	(0.025) (0.188)	0.137** (2.175)
Family size	(+)	-0.0039 (-0.150)	0.0098 (0.482)
Raksi	(+)	0.246 (1.233)	
Brahmin	(+)		0.0025 (0.291)
Possession of an improved stove	(-)	-0.331* (-1.806)	-0.260 (-1.160)
log likelihood function		-347	-442
degrees of freedom		230	246

[a] Parenthetical expressions are expected signs and asymptotic t statistics. ***, **, and * indicate statistical significance at the 0.01, 0.05, and 0.10 levels, respectively. These are log-log 2SLS regressions. Therefore, the estimated coefficients are elasticities.

Market Fuelwood Expenditures

Table 7.5 shows our equations for the household expenditures for market fuelwood, eq. (7.14b). The log likelihood functions indicate satisfying equations. Because these are linear equations, we estimated elasticities at the

Table 7.5. Market Fuelwood Expenditures[a]

Explanatory variables		Mid-hills	Tarai
Constant		-12.207 (-0.176)	-131.14* (-2.343)
Estimated wage	(?)	-2.942*** (-3.720) [-0.73]	5.682 (0.205) [1.01]
Fuelwood price	(?)	-183.63*** (-3.512) [-0.47]	152.86*** (4.664) [0.79]
Forest area	(-)	0.00055** (1.315) [0.20]	
Forest volume	(-)		-0.0143*** (-2.262) [-0.69]
Distance to nearest trail	(-)	-5.153 (-0.137) [-0.03]	
Distance to nearest road	(-)		0.899*** (3.320) [0.20]
Landholdings	(+)	0.102 (0.137) [0.0005]	2.473 (1.450) [0.07]
Family size	(+)	5.883* (1.813) [0.25]	-1.003 (-0.360) [-0.08]
Raksi	(+)	-50.766* (-1.776)	
Possession of an improved stove	(-)	16.628 (0.729)	39.960 (1.114)
log likelihood function degrees of freedom		-491 231	-959 247

[a] Parenthetical expressions are expected signs and asymptotic t statistics. ***, **, and * indicate statistical significance at the 0.01, 0.05, and 0.10 levels, respectively. These are linear Tobit 2SLS regressions. Terms in brackets are elasticities calculated at mean observed values.

regional mean values. The terms in brackets in table 7.5 record these elasticities.

Three of the four crucial wage and price coefficients are significant. The signs on these coefficients indicate tell us a great deal about the differences between hill and tarai households. Tarai households are demand price inelastic.[2] They are also inelastic respondents to changes in their implicit fuelwood collection wages. Hill households are both price and wage elastic, but more wage responsive than price responsive. Apparently, higher implicit wages attract labor to collect fuelwood for internal household use in the hills—and sharply decrease market expenditures on fuelwood. Prices have a less elastic effect on market expenditures in the hills, and neither wages nor prices have much effect on tarai expenditures.

The land variable performs like a proxy for income or wealth in both regions. Larger families (which are probably wealthier as well) spend significantly more on market fuelwood in the hills. Our family size coefficients are further support for the argument that fuelwood is an important component of hill household budgets but not such an important component of tarai budgets. Larger families, and larger incomes, are likely to make a difference in expenditures on items that are important to household budgets, items like fuelwood in the hills. The effects of larger families, and larger incomes will be less important for small budget items. Therefore, we are not surprised by the insignificant coefficient on family size in the tarai.

To this point, our findings for market fuelwood expenditures are consistent with the findings in table 7.4 for households that collect fuelwood. Unlike collecting households, however, households that purchase market fuelwood do not treat improved stoves as technological substitutes for fuelwood. Observations of household behavior with respect to improved stoves have confounded previous researchers (e.g., Amacher et al. 1993). We will return to discuss substitution again when we are in a position to comprehensively review all of our observations together.

Fuelwood Supplied to the Market

Table 7.6 shows our estimates of household fuelwood supplies to the market, eq. (7.14c). The log likelihood functions indicate acceptable equations and all eight coefficients for which we can establish expectations have the anticipated signs. Because these are log-log equations, the estimated coefficients are also elasticities.

Table 7.6. Fuelwood Supplied to the Market[a]

Explanatory variables		Mid-hills	Tarai
Constant		37.132** (3.477)	-14.350*** (-3.642)
Estimated wage	(+)	6.810*** (2.700)	0.547 (1.095)
Fuelwood price	(+)	2.987*** (3.399)	0.360*** (2.589)
Forest volume	(+)		1.569*** (3.516)
Forest area	(+)	0.449*** (3.932)	
Distance to nearest trail	(-)	-0.978** (-2.289)	
Distance to nearest road	(-)		-0.118 (-1.113)
Landholdings	(?)	0.479 (1.173)	-0.0916 (-0.592)
log likelihood function		-419	-313
degrees of freedom		235	249

[a] Parenthetical expressions are expected signs and asymptotic t statistics. ***, **, and * indicate statistical significance at the 0.01, 0.05, and 0.10 levels, respectively. These are log-log Tobit 2SLS regressions. Therefore, the coefficients are elasticities.

Market supply in the hills is both wage and price elastic—but it is much more wage elastic. The anticipated negative sign on resource access reconfirms our prior observations (tables 7.4 and 7.5) that access is a critical determinant of household activity, particularly in the hills where resource scarcity seems to have a more important effect on household behavior.

Tarai market supply is also both wage and price inelastic. Tarai supply is elastic with respect to the stock of standing resource remaining on the common lands, but inelastic with respect to the accessibility of this resource to the market. Private lands in the tarai perform more like an indicator of household wealth than of production or supply—much as they did in our tarai production and expenditure equations. What fuelwood is grown on private land is consumed in the growers' household. Apparently, prices are not yet sufficient to induce landowners to grow enough fuelwood to affect market supply across the breadth of the entire region. What fuelwood is grown on private land is consumed in the growers' household. Apparently wealthier tarai households may grow some of their own fuelwood but they are more like to purchase market fuelwood and less likely to be involved in supplying fuelwood to the market.

Combustible Agricultural Residues

Agricultural residues are generally considered to be a fuelwood substitute, although Nepali households consider them an inferior fuel and prefer not to burn them when fuelwood is available. Residues also have important alternative uses as a livestock food supplement and for their contribution to agricultural sustainability when they are returned to the soil. Hill households in particular consume substantial quantities of residues (table 7.1).

Table 7.7 shows our estimates of household residue consumption, eq. (7.14d)—and our check on the importance of residues as a fuelwood substitute. The log likelihood functions indicate acceptable equations but the signs on many coefficients in both the hill and the tarai equations are unconvincing. Because residues are not traded, we have no observations on residue prices. Fuelwood prices are our only market indicator and positive fuelwood price coefficients would indicate that residues are fuelwood substitutes. Our coefficients do not support this contention.

The negative signs on the land coefficients indicate, once more, that land apparently acts more like a proxy for household wealth than a source of residues. That is, larger (and wealthier) households consume less residues. This effect is particularly true in the hills where fuelwood is a more important draw on the household budget. This negative sign reaffirms our expectation that residues are an inferior good. The negative sign on improved stoves indicates substitution in consumption—and access to the forest stock (but not the level of the stock itself) performs like a substitute in production.

In sum, we cannot conclude that residues are an important fuelwood substitute. This is what we would expect where fuelwood is not in severe shortage or where factors other than fuelwood (e.g., livestock and the alternative role of residues in maintaining soil productivity) have greater substitution impacts on residue consumption. Perhaps residues are a fuelwood substitute for only a minority of Nepali households, and perhaps most households perceive that crop residues are an essentially free joint output of agricultural production with no critical alternative household uses.

Implications for Scarcity and Development Opportunity

Our results suggest a "story" for the over-riding policy concerns with deforestation and fuelwood scarcity that emerges more clearly from a summary of the estimated elasticities. We have argued that fuelwood scarcity has not yet become a critical economic factor for many Nepali households. It is important for others, particularly for collecting households from some hill districts.

Table 7.7. Consumption of Combustible Agricultural Residues[a]

Explanatory variables		Mid-hills	Tarai
Constant		-1025.50*	184.87***
		(1.688)	(2.545)
Estimated wage	(−)	28.059**	-269.14**
		(2.245)	(−2.193)
Fuelwood price	(+)	-400.14*	0.225
		(−1.713)	(1.235)
Forest area	(−)		-0.00074*
			(−1.618)
Forest volume	(−)	0.0029	
		(0.514)	
Distance to nearest trail	(+)	39.794	
		(1.031)	
Distance to nearest town	(+)		0.616*
			(1.846)
Landholdings	(?)	-8.045**	-0.588
		(−2.245)	(−0.411)
Family size	(?)	-0.204	3.152
		(−1.186)	(1.064)
Possession of an improved stove	(−)	-300.19***	74.317**
		(−2.777)	(2.285)
Animal units	(?)	-18.061	2.873*
		(−1.124)	(1.692)
District population	(?)	0.0028** *	
		(3.529)	
log likelihood function		-189	-169
degrees of freedom		232	246

[a] Parenthetical expressions are expected signs and asymptotic t statistics. ***, **, and * indicate statistical significance at the 0.01, 0.05, and 0.10 levels, respectively. These are linear 2SLS regressions.

Regional Scarcity

Total resource stocks are generally lower in the hills than the tarai. Yet hill households generally consume more fuelwood and also more agricultural residues. Does this mean that fuelwood scarcity is not a problem? Does the evidence that some tarai households have begun producing fuelwood on their own scarce agricultural lands mean that fuelwood availability is critical for them? Table 7.8, which reviews our household elasticities, suggests some answers.

Of those households that participate in market exchange, hill households are much more demand responsive, and somewhat more supply responsive than tarai households. Tarai market demand in particular is unresponsive to market change. Hill households respond to higher prices by sharply decreasing their market fuelwood expenditures—and they respond to the higher implicit wages that accompany higher prices by increasing their own collection activity in an effort to offset the decrease in their market expenditures. Tarai households hardly adjust either their market expenditures or their collection.

This suggests that deforestation and fuelwood scarcity have greater potential impacts on household behavior in the hills. We might expect that labor reallocation in order to increase fuelwood collection would cause the more responsive hill households to search for labor-saving opportunities in other activities and to display willingness to substitute away from fuelwood collected on the forested commons, with substitutes like alternative fuels (e.g., residues) and alternative technologies (e.g., improved stoves), and with fuelwood production on their own lands. The evidence in not convincing on any of these. Our regressions provide no evidence of market demand substitution for either agricultural residues or improved stoves in either the hills or the tarai. And, in previous analyses of hill households, Cooke (1996) observed that slack season production opportunities exist and Amacher et al. (1993) observed that the fuelwood labor burden may be less than it sometimes seems because collection time is a joint input to activities like child care and socialization as well as to fuelwood production.

Households in both regions are more market supply responsive. Hill market supply is exceedingly wage responsive—which is consistent with the greater wage responsiveness of hill households in market demand, and also with our knowledge that the average hill household is poorer than the average tarai household (see table 7.1). Tarai supplies are notably responsive to the level of resource stock, but not to resource access. Undoubtedly, this reflects the easier terrain of the tarai.

Table 7.8. Household Elasticities, Revisited

			Mid-hills	Tarai
Household expenditure elasticities with respect to				
Estimated wage		(?)	-0.73*	1.01
Price^2		(?)	-0.47*	0.79*
Land (proxy for wealth)		(+)	0.0005	0.07
Family size		(+)	0.25*	-0.08
Improved stoves		(-)	+	+
Household supply elasticities with respect to				
Estimated wage		(+)	6.81*	0.55
Price		(+)	2.88*	0.36*
Resource stock (area/volume)		(+)	0.45* (area)	1.57* (volume)
Improved resources access/ decreased market access		(-)	-0.98*	-0.12
Household collection elasticities with respect to				
Estimated wage		(+)	1.33*	0.09
Price		(+)	1.57*	0.71
Improved stoves		(-)	-0.33*	-0.26
Resource stock (area)		(+)	0.45* (area)	0.26 (area)
Improved resource access/ decreased market access		(+)	0.22*	-0.06
Landholding		(+)	0.03	0.14*
Family size		(+)	-0.004	0.01

[a] The demand price elasticities are -1.47 for the hills and -0.21 for the tarai.
* indicates elasticities derived from coefficients that are statistically significant at least at the 0.10 level. Supply elasticities are not strictly comparable across the two regions because the hill measure of resource stock is forest area while the tarai measure is forest volume.

We cannot conclude, however, that these supply responses are the only household responses to increasing fuelwood scarcity. Households in both regions collect fuelwood for their own consumption, and collection is more responsive than market demand to the conditions of fuelwood scarcity.

Increasing market prices induce a compensating increase in the implicit wage for fuelwood collection and in fuelwood collection itself, especially in Nepal's hills. As market prices increase, some households, especially in the hills, withdraw from market activity altogether and collect the household's entire consumption. Collecting households in the hills show the first convincing evidence of substitution. They do substitute improved stoves for scarcer fuelwood. Furthermore, for collecting households and especially in the hills, we see that the volume of fuelwood collected is significantly (although inelastically) responsive to the level of resource stock and to its accessibility. Therefore, the availability of the hill stocks constrains the final level of fuelwood consumption by these households.

For collecting households in the tarai, we see that collection bears a more reliable relationship with fuelwood production on private lands than with the household's implicit wage for fuelwood collection. Therefore, we might anticipate that, for those tarai households who do modify their demand behavior, the modification is production (collection) oriented; and land and resource stocks, rather than labor, are the substitutes of choice. Nevertheless, it is clear that total tarai collection, like tarai market demand, is relatively unresponsive to market change.

Overall, these observations fit the general economic description for a good that consumes only a small part of the budget of most households. That is, higher prices would not be expected to substantially alter demand for a good which has little effect on the household budget. In our case, higher fuelwood prices do alter market demand and supply in the expected directions but apparently, for many hill or tarai households, the adjustments in market fuelwood consumption are insufficient to substantially modify overall household behavior.

Our family size variable lends conviction to this explanation. Larger families should consume more fuelwood—unless households have the means to provide a plentiful level of consumption regardless of family size. This seems to be the case with tarai demand. On the other hand, market expenditure in the hills is more responsive to family size. Fuelwood probably is a more important budgetary item for hill households but, even in the hills, the market demand response is small. So—should we conclude that fuelwood is not scarce and deforestation has not created a fuelwood problem? Not yet.

We have seen that some households do alter their demand behavior by refraining from market purchases altogether and collecting their entire consumption. These households are a majority of all households in our sample and fuelwood scarcity is a burden for them. It does cause meaningful adjustments in their behavior that are confirmed by their willingness to substitute improved stoves and agricultural residues for fuelwood. We observed this shift notably in the hills and we note that those hill households that make the shift are poorer households (for which the market price adjustment would have a greater budgetary impact) with smaller families (for which an increase in fuelwood collection time would have a greater affect on total labor allocation). These collecting households, and the districts where they comprise a large share of all households, could be good target populations for fuelwood intervention policies and programs.

More Localized Scarcity

The policy problem is to predict the districts, especially in the hills, where localized fuelwood scarcity is most important, and to anticipate reasonable policy responses. Ideally, we would have evidence of the price elasticities for potential fuelwood substitutes and evidence of the cost differentials that would induce households to adopt lower opportunity cost substitutes for fuelwood. In the absence of the cost information, we might anticipate that more elastic responses to fuelwood prices and the implicit wages would indicate a significant impact on household budgets and a greater likelihood that households will adopt fuelwood substitutes. In the event that collection is wage and price elastic, current substitution behavior might also be indicative. More elastic substitution in consumption (improved stoves and agricultural residues) might indicate a potential willingness to substitute alternatives like market fuels and further improvements in stove technologies. Similarly, more elastic associations with alternative fuelwood sources (community plantations and private resource stocks) might suggest greater household opportunity to save labor by substituting new sources and a better opportunity for policy interventions like tree planting on the common lands, like nurseries to provide seedlings for private lands, and like forest extension advice for landowners and communities.

We can create some of this evidence by re-evaluating the elasticities for those collecting households in the hills that our previous analysis showed to be most wage and price. The elasticities we report will be the regional collection elasticities evaluated at the mean values reported for our survey households in each district. Table 7.9 shows our results for thirteen hill districts arrayed in their geographic sequence from west to east. The improved stove column is empty because we cannot calculate elasticities for qualitative variables. The residue column is empty because our hill residue equation does not support the contention of fuelwood substitution. Leaving these columns in the table reminds us that there is a possibility for substitution opportunities in consumption, a possibility that may be more important for fuelwood consuming households in some other parts of the world.

Kathmandu emerges from table 7.9 as the district with highest fuelwood prices and most elastic wages and prices. Prices are high enough that Kathmandu households are very responsive to increasing fuelwood scarcity. We do not observe unusual responses by local households to the forest stock or private planting opportunities because Kathmandu is a developed district, and such opportunities are sparse in this relatively urban district. We would anticipate greater substitution in consumption—for improved stoves or for alternative fuels. In fact, kerosene use was widespread in Kathmandu at the time of our survey. (Kerosene use was excluded from the survey because it was largely unavailable in other districts of Nepal's hills.)

There are no uniform conclusions to be drawn about the remaining districts where households are very wage and price responsive. Each district is different. Some individual cases may be instructive. Rukum and Argha Khanchi are very wage and price responsive, and Argha Khanchi is a high price district. Both districts are in the less accessible western hills where we might anticipate that improved roads and trails could make a difference. Substitution in consumption is probably not a realistic alternative in these districts until the road system is improved because most consumption substitutes (improved stoves, kerosene, LPG) require importation from external sources.

Palpa is both wage and price responsive. It is an intensively cultivated district. Therefore, it is not surprising that Palpa's forest stock per capita is low and that this district is among the more responsive to private tree planting opportunities.

Kaski and Gorkha are among the more wage responsive districts. These are more populated districts with better road and trail systems. Therefore, the market fuelwood and household wage opportunities are greater in response to increasing scarcity. In Kaski in particular, resource access seems to be an important determinant of fuelwood consumption. Kaski and Gorkha would be good candidates for further examinations of the substitution opportunities in consumption.

Dhading is a well-developed agricultural district near the Kathmandu population center. Dhading households often provide fuelwood for the Kathmandu market. Therefore, it should not be surprising that they are especially wage responsive and it should not be surprising that these households are unusually responsive to the private tree growing opportunities that their own agricultural lands and the Kathmandu market provide.

No other district is both wage and price responsive. For example, Ramechhap and Llam are in the relatively better roaded eastern part of the country. Therefore, their households are among the more responsive to differences in resource access. Ramechhap forests are unusually poorly stocked. Therefore, while fuelwood prices remain low in this district, it is not

Table 7.9. Mid-Hill District Mean Prices and Collection Elasticities

District	Price (r/kg)	est'd wage	price	imp'd stoves	residues	resource stock	resource access	private land
					Collection elasticities			
Rukum	0.40	1.63	1.08			0.91	0.14	0.014
Baglung	0.59	0.59	0.57			0.24	0.068	0.012
Gulmi	0.39	1.04	0.72			0.20	0.085	0.019
Argha Khanchi	0.74	1.22	1.30			0.32	0.014	0.023
Palpa	0.55	2.01	1.81			0.47	0.026	0.040
Kaski	0.43	1.45	1.16			0.58	0.34	0.025
Lamjung	0.38	1.22	0.94			0.35	0.71	0.016
Gorkha	0.41	1.68	1.17			0.69	0.14	0.024
Dhading	0.37	3.84	2.89			1.45	2.00	0.066
Kathmandu	1.64	3.43	15.06			0.24	0.37	0.010
Sindhupalchok	0.47	1.27	0.92			0.46	0.16	0.033
Ramechhap	0.20	1.47	0.36			0.17	0.33	0.058
Laim	0.61	0.54	0.73			0.25	0.43	0.026

surprising that a few households are responsive to private reforestation opportunities.

In sum, only one general observation emerges from table 7.9: the futility of broad general conclusions about either fuelwood scarcity or the selection of preferred interventions to assist local households in addressing it. Even the presence of higher prices is not a perfect indicator of responsive households and districts. The evidence from table 7.9 is consistent with our prior observation that fuelwood is not a generally scarce resource throughout Nepal—but it is scarce for some households in some districts. Those districts and their households have their own specialized characteristics. Therefore, where important evidence of scarcity exists, successful program and policy interventions must be designed to match those local characteristics. This is an important finding for policy because many international donor agencies have designed interventions to substitute for fuelwood consumption and to diminish deforestation in Nepal's hills. Our evidence suggests that these interventions, to be successful, must be targeted to select districts and to collecting households. These households are not the largest consumers of fuelwood and they generally are not the better-off households from which local leaders tend to emerge. These households are difficult targets, indeed, for policy intervention and behavior modification.

Summary and Policy Conclusions

We set out to examine differences in resource scarcity across broad regions, and to identify the areas within a region most likely to respond to opportunities for new social forestry interventions. We used a model that incorporates household fuelwood collection, as well as market fuelwood expenditures and supply, to demonstrate our analysis across the mid-hill and tarai regions of Nepal.

Our most important observation is that market information alone is insufficient. Market evidence does not reflect household collection behavior. Therefore, market demand and supply information on consumer behavior can vary greatly from actual household behavior when many households collect fuelwood for their own internal use and when their household production and consumption decisions are inseparable.

An example of the differences between market evidence and evidence from the full range of household activities illustrates this point. Consider the expenditure price and supply price elasticities derived from either market evidence alone or the full range of household evidence.

	Mid-hills	Tarai
Expenditure price		
Market evidence only	1.52	0.04
Full household evidence	-0.47	0.79
Supply price		
Market evidence only	0.39	0.73
Full household evidence	2.99	0.36

Clearly, market estimates widely misrepresent household behavior in either of our two survey regions.[3] Moreover, this comparison also suggests a broader implication. It suggests that we should anticipate similar divergence between household and market evidence for many other forest-based activities of subsistence households—in Nepal and in other regions around the world.

NOTES

1. Ministry of Forestry, Kathmandu; Forestry Department, Virginia Tech; and Centre for International Forestry Research, respectively. Our analysis benefits from Gunnar Kohlin's most thoughtful comments.
2. Expenditure elasticities are 1+ (demand price elasticity). Therefore, the demand price elasticities are -0.21 and -1.47 for the tarai and hills, respectively.
3. The household elasticities are from table 7.8. The market elasticities were estimated independently. Market expenditure is a linear Tobit function of own price, improved stoves, land, and family size; and market supply is a log-log Tobit function of own price, forest area or volume, and forest access. These functional forms are identical with our household estimations. The RHS variables are similar except for the inclusion of the estimated wage term in our household supply functions. Therefore, the market and household functions are as comparable as possible. Both market equations performed satisfactorily and the signs on their coefficients followed expectations. Only the hill expenditure price coefficient was significant in the market equations, while all four price terms were significant in the household equations.

REFERENCES

Amacher, G. S., W. F. Hyde and K. R. Kanel. 1996. "Household fuelwood demand and supply in Nepal: choice between cash outlays and labor opportunity." *World Development.* 24(11):1725–1736.

Amacher, G. S., W. F. Hyde, and B. R. Joshee. 1993. "Joint production and consumption in traditional households: fuelwood and agricultural residues in two districts of Nepal." *Journal of Development Studies.* 30(1):206–25.

Bluffstone, R. 1995. "The effect of labor markets on deforestation in developing countries under open access: an example from rural Nepal." *Journal of Environmental Economics and Management.* 29(1):42–63.

Cooke, P. A. 1998. "The effect of environmental good scarcity on own-farm labor allocation: the case of agricultural households in rural Nepal." *Economic Development and Cultural Change.* (forthcoming).

Cooke, P. A. 1996. "The effect of environmental degradation and seasonality on the collection and consumption of environmental goods by rural Nepali households." Unpublished manuscript, Department of Development Economics, Free University, Amsterdam.

Goetz, S. 1992. "A selectivity model of household food marketing in Sub-Saharan Africa." *American Journal of Agricultural Economics.* May.

Kanel, K. 1996. "Farm forestry in the Tarai of Nepal: a policy perspective," *Banko Janakari.* 6(1):11–19.

Shaikh, A., with K. Kanel, D. Koirala, and S. Pandy. 1989. "Forest products marketing systems in Nepal." Unpublished manuscript, US AID Mission to Nepal, Kathmandu.

Singh, I., L. Squire, and J. Strauss. 1986. "The basic model: theory, empirical results, and policy conclusions," in I. Singh, L. Squire, and J. Strauss (eds.), *Agricultural Household Models.* Baltimore: Johns Hopkins University Press, pp. 39–69.

Thornton, J. 1994. "Estimating the choice behavior of self-employed business proprietors: an application to dairy farming," *Southern Economic Journal* 87(4):579–95.

Part 5
Secure Rights

Our analyses to this point have focused on households, their resources, and their behavior. Our analyses, and much of economic development literature, assume that households have secure rights to the resources they use. Secure rights describes the norm for most assets, particularly those in scarce supply, and it fits the personal experience of most analysts. In fact, secure rights are an increasingly appropriate description in social forestry—as farmers and farm households increasingly use their own labor and capital resources to plant and manage trees on their own land.

Nevertheless, forestry's position at the geographic margin of developed land uses and the frequent availability of open access forest substitutes mean that uncertain and insecure rights are common in forestry. Chapters 8 and 9 address this issue and its implications for social forestry.

Chapter 8 traces the general implications of changing forest values for secure tenure of the forest resource. The chapter's special reference is the Philippines, and the current Philippine program to transfer many state responsibilities for forest land to local communities. Rising values for the forest resources mean increasing incentives to establish secure long-term claims on trees and forest land. Lower costs of local enforcement reinforce these incentives. The problems are to identify where prices have risen sufficiently to justify changes in tenure, to identify where these changes are about to occur, and to simplify the transition from state to private or community rights for the forest resource. Many of the truly exciting opportunities in social forestry today lie in a fair sorting of these problems. Unlike our other chapters, this chapter is conceptual not empirical. The problem, however, is a real problem for the Philippine Bureau of Forest Development and for many other forest ministries. Clear organization of the concepts at stake should help arrange a lasting solution.

Chapter 9 addresses the related issue of uncertainty, and reinforces a point raised in chapter 1 and again in chapter 8 that changing government policies create local uncertainty and operate against any long-term incentives—including incentives for private investments in forestry. Two regions of China provide the special case for chapter 9. The first has a recent history of consistent policies increasing local responsibility for land use decisions. The farm forest response in this region has been phenomenal, with significant increases in both timber harvests and reforestation. Government policy in the second region has changed course several times and the farm forest response is much less satisfying. Farmers with rights to trees in the second region harvest as rapidly as they obtain the rights, and they harvest without replacement precisely because they cannot depend on the rights to their reforestation investments. Our conclusion is that a stable policy environment is a prerequisite for long-term investments. Policy uncertainty is a form of removing secure rights to the land. Without a predictable policy environment, farmers and local communities risk losing the rights to any investment they may make in their land or trees. They understand this and refrain from investing. Sustainable forestry cannot exist in the absence of a stable policy environment.

CHAPTER 8

Secure Forest Tenure, Community Management, and Deforestation: A Philippine Policy Application

Marcelino Dalmacio, Ernesto S. Guiang, Bruce Harker, and William F. Hyde[1]

Foresters have a dilemma. Forests are extensive resources with dispersed pockets of high commercial value and other pockets of great environmental risk. Protecting these resources is difficult because it is difficult to exclude unauthorized users from the forest—and most forests are subject to illegal logging and other incidents of forest trespass. The state generally owns the forests and the public holds foresters responsible for both the high-value sites and for the extensive regions of occasional trespass. The foresters' dilemma is compounded by the contrast between national and local values for the forest. National concerns with deforestation, biodiversity and the unique characteristics of the natural environment compete with the local importance of forests as sources of land for the agricultural development of expanding upland populations.

This dilemma causes many foresters to be sympathetic with the idea of community-based forest operations—but with strict controls on the community operations and careful collection of fees for wood resources. It seems to describe the situation in the Philippines, but comparable concerns arise for government forestry agencies around the world, in developed and developing countries alike. In the Philippines this dilemma describes the Department of Environment and Natural Resources' (DENR) justification for its Community Based Forest Management Program (CBFM) and it seems to underlie DENR's evolving position on Industrial Forest Management Agreements (IFMA). CBFM places public forest land under local community management. IFMA takes public land from canceled Timber Licensing Agreements and encourages its sustainable management by commercial timber enterprises. Both intend to ease the government's administrative responsibilities and increase private initiative while maintaining DENR oversight to protect the environment and recover public revenues. CBFM in particular fits within our perspective of social forestry in this book.

The objective of this chapter is to examine CBFM and IFMA within the general context of efficient forest management. We will look for ways to accomplish agency, community, and commercial forestry objectives while decreasing costs for each of these institutions. The key is to focus DENR oversight on the resources it protects—land and forests, rather than logs—and on the specialized cases across the forest landscape that pose significant environmental risks or promise large revenues for the government treasury, while reducing DENR's management costs in the more general and lower-value cases where either local communities or a commercial enterprise can do a better job.

Our chapter begins by reviewing the underlying principles of forest development and by identifying the potentials for sustainable management and the collection of forest revenues. We will also consider the administrative costs required to protect the environment and those revenues. We will see that transferring property rights from DENR to local communities or private managers will not guarantee sustainable management and that plantation forests are not always a viable substitute for resource extraction from natural forests. That is, local communities and local private management can improve the conditions for sustainable land management and even for forest plantations, but it will never be economically efficient to manage all forests sustainably.

The latter part of the chapter will examine CBFM and IFMA in the Philippines within the context of these general principles. Community management and private forest enterprise have much to recommend them, but the current level of DENR oversight interferes with both. Our main conclusion is that redirecting DENR's financial and human resources to focus on the most critical environmental tasks and on those higher-value timber opportunities which foresters are better trained to manage will improve local economies, increase the potential for sustainable management, and increase the net collection of government revenues. The same principles could be applied to government forest management around the world.

Principles of Forest Development

The most general principles of forest development are contained in a simple diagram of the agricultural and forest landscape. Consider agricultural land first. The value of agricultural land is a function of the net farmgate price of agricultural products—which is greatest near the local market at point A in figure 8.1. Agricultural land value declines with increasing distance or decreasing access to the market as described by the horizontal axis. The function c_r represents the cost of establishing and maintaining secure rights to this land. This function increases as public infrastructure and effective control decline with distance from the market. The cost of obtaining legal title may remain constant with distance but the cost of excluding trespassers increases dramatically until no number of forest guards can fully exclude squatters, illegal loggers and other local users of the land from remote locations.

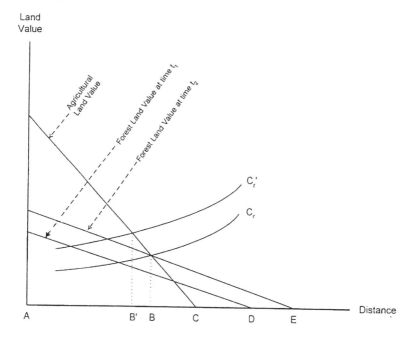

Figure 8.1. The landscape of forest development

The functions explaining agricultural land value and the cost of secure property rights intersect at point B. Farmers will manage land between points A and B for sustainable agricultural activities. They will use land between points B and C (where agricultural value declines to zero) as an open access resource to be exploited for short-term advantage whenever possible. They will harvest natural crops in this region, crops like fodder for their animals, native fruits, and fugitive resources like wildlife, but they will not invest even in modest improvements in the region between points B and C because the costs of protecting their investments in this region are greater than the investments are worth. Their use of this open access region is unsustainable.

The mature natural forest at the fringe of agricultural development at point B has an initial negative value because it gets in the way of agricultural production and its removal is costly. When the standing forest begins to take on positive value, it must be worth less than agricultural land and the function describing forest value must intersect the horizontal axis at some distance beyond point B. Market demand justifies the harvest of forest products at this point, and it continues to justify harvesting until the forest frontier eventually shifts out to some point like D. At this point the market price of forest products just covers the cost of their removal and delivery to the market. The *in situ* resource price (the stumpage value for timber, for example) is zero, and the value of forest land at D is also zero. The region of unsustainable open access activities now extends from point B out to point D. The costs of obtaining and maintaining property rights insure that this region will remain an open access resource. Some trees will re-grow naturally in this degraded region and the region's natural resources will be exploited in periodic "pulse" harvests as soon as they grow to their minimum economic value. The region beyond point D remains an uneconomic and unexploited natural forest residual because the costs of its exploitation exceed the value of the products removed.

This story conforms with the common description of any initial settlement, including modern settlement by migrants to the Philippine uplands. It also conforms with the pattern of land use in stable upland settlements, and it conforms with our image of commercial logging in the Philippines for the last forty years. We described the pattern of initial settlement. Consider the other two cases of stable upland communities and commercial logging.

Farmers in stable upland communities make long-term conservation investments in agricultural technologies like terraces on land where their property rights are secure, and sometimes they walk long distances to the natural forest for products like fruit and rattan. Some use the open access areas between their agricultural plots and the forest to graze their livestock, a traditional pulse activity, and many continue to collect fuelwood from the degraded open access areas as soon as the scattered woody matter grows to a burnable size.

The forest first slows the rate of agricultural development but later, as timber becomes a valuable commodity, loggers begin to clear an area beyond the agricultural settlement. As timber becomes increasingly valuable, loggers clear areas ever-farther from their prime markets. They "cut and run" and leave a "no-man's land" behind them. They always harvest to the point where the full resource value is dissipated in harvest and access costs because if they left "value on the stump" another logger would take it.

Closer examination of the cost of securing the property rights points to another important case. The costs of protecting these rights are greater for absentee landowners than for local landowners. This suggests an absentee landowner cost function like c_r' where land to the right of the new point B' is effectively open access regardless of any legal registration to an absentee owner. Transferring land in the region between points B' and B to lower-cost local owners would encourage its sustainable management. Of course, DENR is an absentee landowner, and this explanation is one rationale for transferring some DENR properties to upland settlers or to lower-cost community forestry operations.[2]

Consumers in our landscape in figure 8.1 eventually must go beyond point D if they continue to demand the products of forest resources. Access costs for forest products (and the product prices) must continue to rise until some later time t_2 when the higher market price equals the backstop cost of tree crop substitutes for forests (including the cost of local property rights c_r). When this price also exceeds the opportunity cost for agricultural land, then trees will begin to compete as successful crops (either plantation forests or agroforestry) somewhere to the left of point B (or point B' for the properties of absentee landowners). Some forest products will continue to be removed from the new fringe of mature natural forest at E. Subsequent to time t_2, plantation forests to the left of point B and pulse harvests from the region between points B and E will satisfy all demands for forest products. No further extraction will occur from the natural forest beyond point E because the access costs associated with ventures into the natural forest would exceed the market price and the alternative costs of plantation production.

This story is well-documented with rigorous economic analyses, and with examples from many cultures and developmental experiences.[3] There are many examples of subsistence farmers in the Philippines who plant trees in agroforestry operations or small plantations, but there are many other examples in other parts of the Philippines where the prices of forest products are not yet sufficient to induce this smallholder activity. This means that some local markets (perhaps like those in most of Palawan) fit the description of our landscape at t_1 with a natural forest fringe at D, while others (perhaps much of Bohol and Cebu, for example) fit the description at t_2 with a natural forest fringe at E and some tree crops to the left of point B. The industrial plantations of PICOP or those other plantations near Butuan City, for example, are evidence that prices in the world markets for commercial forest products are a sufficient private incentive to support our t_2 characterization of them.

Modifications of the Principles

Our diagrammatic presentation is simplified because it shows an unchanging agricultural demand and a single price for forest products, and because it overlooks stock effects. Growing upland populations and increasing food demands would push the agricultural land value function to the right over time. This would shift points B and C to the right. Additional forest products, each with their own prices and forest value functions, would have their own land value gradients and their own points D and E—and local inhabitants would travel different distances into the standing natural forest to extract different products. The general characterization of the problem, however, remains the same in each case.

Stock effects measure the forest's value as a standing resource. If we anticipate that either biological growth or changes in forest product prices will increase the value of standing timber over time, then current market prices will underestimate the true long-run value of the standing forest. The forest frontier set at either D or E set according to current market value would be too far to the right and the current level of logging would be excessive from a social perspective. On the other hand, this problem should not concern us too much because the natural forests at the fringe of economic activity are mature. Their growth rates are insignificant. Moreover, virtually all world natural resource prices for which we have extensive records have declined (in inflation-free real terms) over the last 200 years. World prices for tropical hardwood logs are probably holding steady, and they may eventually decline as well. Therefore, the stock effects are probably small and they may be negative. Negative stock effects (due to decreasing real prices) would redouble the incentive for loggers to harvest as far as they can today because the values they pursue may be lower tomorrow.

Fundamental Conclusions

The fundamental conclusions of this analysis should be clear: 1) An expanding region of deforestation causes forest product prices to rise, and rising prices eventually induce investments in agroforestry and commercial forest plantations and bring an end to deforestation and the exploitation of the remaining natural forest. This insight provides a bright ray of hope for global concerns for tropical deforestation and for our own desire to protect some of the Philippines' remaining stock of native trees. 2) There is an important and extensive region in which open access and unsustainable management is efficient because the costs of defending secure rights to this region exceed the fundamental values at risk. Environmental and other non-market forest values that originate in this region will require the specific attention of the public managers who work for ministries like the Philippine's DENR.

This second point bears repeating. People will use the land between points B or B' and D or E, but they will treat it as an open access resource. They will not manage it sustainably because sustainability implies long-term decision making and protecting the land for future uses, but the costs of protection exceed the values at risk. Legal or administrative action can shift some responsibility from DENR to the local community or local land manager (thereby shifting the cost function from c_r' to c_r and shifting point B' toward point B). This is a desirable action, but neither legal nor administrative action alone can alter the condition of unsustainable land use altogether. Some land will remain between points B or B' and D or E and no private investor will manage this land sustainably.

Moreover, at the frontier of natural forests, government regulation in any form is expensive (the difference between the horizontal axis at D or E and c_r') and it protects a resource that has no market value. This means that pure economic rents seldom exist because some alert entrepreneur would have harvested to the margin at point D or E, and extracted the rents in a previous period. The entrepreneur would have been there legally if there were no government restrictions. If legal or administrative restrictions did exist, but the enforcing agency could not afford the full costs of complete enforcement implied by c_r', then some entrepreneur would have found a way to be there illegally. Clearly, this shows us that it is easier to write policy than to enforce it, and that the gains from enforcing many forest policies do not justify their costs.

Our two fundamental conclusions are valid for commercial timber. They are also valid for most other forest resource values because most forest values are represented in local markets even where the predominant economy is subsistence based. They are true in Canada, the US, and Siberia, and they are true in the greater reaches of Manila in, for example, the forested areas of the province of Laguna, and they are also true in the sitsos of South Cotabato. The important difference will be that forest product prices have risen to the level that justifies tree crops in some regions but will not have risen to this level in others.

The critical implications for public policy are:

- to simplify the pursuit of forest charges for the government treasury, and

- to examine the policy environment for ways to shift the property right functions in figure 8.1 and the patterns of land use in ways that will improve the incentives for long-term sustainable management and satisfy social and environmental objectives.

These are the challenges—and they should be the guiding principles of forest policy design. They are the spirit of the Philippine support for Community Based Forest Management and, perhaps, for Industrial Forest Management Agreements. They deserve amplification in order to demonstrate the reasonable policy outcomes—for their Philippine applications, and also for similar applications almost anywhere in the world.

CBFM is designed for the many cases of smaller and local forest values where DENR has tried but been unable to enforce its preferred actions. The expectation is that local operators with site specific knowledge and lower enforcement costs can do better. Furthermore, the shift from DENR to local responsibility will ensure an increase in the sustainable land use base comparable to region B'-B in figure 8.1. This is a desirable result. Whether the local forest product price is sufficient to induce tree planting and long-term forest management is an empirical question. Some of the increase in sustainable management will convert previously degraded forest to sustainable agricultural uses. Some may become sustainable tree crops. Both are sustainable. Therefore, both are environmental improvements, but some open access and unsustainable forest exploitation will occur under any scenario. The government and the Philippine public must accept this.

The larger industrial operations that produce higher-value forest products for the broader economy are a related case that will benefit from the closer examination of forest rent in the next section of this chapter. It should be clear that the government's reasonable objective for these operations can be explained as reducing its oversight costs (from c_r' toward c_r in figure 8.1) while still ensuring good environmental performance. Once more, this action will ensure an increase in the sustainable land use base comparable to region B'-B, but whether the local forest product price is sufficient to induce plantation management is an empirical question that will vary from location to location and commercial operation to commercial operation. Legal requirements for reforestation will not make it happen. The recent Philippine experience of unsuccessful reforestation subsidies for Timber License Agreements is clear evidence of this point. Once more, some unsustainable open access forest exploitation will occur under any scenario, and government and the Philippine public must accept it—for CBFM, for IFMA, or for any other management arrangement.

The best way to address the less-satisfying conclusion that some land will remain subject to unsustainable exploitation is to remind ourselves why we object it—and to address the objection. We object because, in some important cases, unsustainable land uses degrade the environment, and environmental degradation affects future uses of this land and its resources, as well as the future use of adjacent lands. In forest environments, the major varieties of environmental degradation are soil erosion (steep slopes, streambanks, and critical watersheds) and loss of biodiversity (old growth forests and rare and endangered species). Fortunately, soils do not erode and biodiversity is not lost in every location where someone extracts forest resources. Fortunately too, foresters are trained ecologists whose basic instincts are to respect these environmental aspects of the forest. Therefore, the solution is for DENR's to identify the critical areas for protection against erosion and loss of biodiversity before the unsustainable extraction activity begins, to restrict extractive activities on these areas or to remove these areas from the land base eligible for extractive activities, and to direct a renewed government focus on these critical areas. No country's forest ministry has the financial resources and the personnel to worry about all areas. It must concentrate its limited resources on the critical areas where the values at risk are greatest.

Forest Rent Revisited

It may be instructive to take a closer look at forest rent, and also at the impact of uncertain policy expectations. An understanding of the potential for rent will be important for assessing reasonable charges for industrial forest operations, and perhaps for some community-based operations. An understanding of the destructive effects that uncertain policy environments have on long-term capital investments will be important in assessing reasonable expectations for sustainable industrial forestry operations.

Our earlier claim that rents are uncommon begs questions about why loggers continue cutting trees, why their profits seem excessive, and under what uncommon circumstances significant rents do arise. Loggers continue cutting because there is sufficient opportunity at the margin to obtain a fair return on capital investments in logging equipment. Excess profits are unlikely, however, as long as there are no barriers that restrict entry to the logging business. New entrants will compete and bid away the excess.

Purchasing the logging equipment, building a road to the site, and delivering the equipment to that site is expensive. (It requires a large capital investment even when the logging itself is done with chainsaws and caribao). Road costs alone generally exceed the value of a single block of timber, and loggers only recover these road costs by harvesting several blocks of timber off the same road. What may seem like a large absolute value for the timber taken from a logging contract often barely reimburses the capital expenditures and leaves very little for a fair return on the capital investment. We know this because the logging business is like the small retail sector or the restaurant business. There are many small entrepreneurs and many failures for every lasting large success.

Indeed, logging operators in the Philippines are too numerous and often too small to show up in government records, and even the sawmill sector has a large component of small entrepreneurs and only temporary successes. For example, sixteen percent of sawmill capacity in the Philippines in 1994 was classified as mini-sawmills and more than one-half of these were inactive (DENR, nd).

Sources of Rent

Nevertheless, examples of large rents do exist, and we can imagine the potential for excessive profits where some exogenous action creates a sharp and rapid, even sudden, shift in the forest land value function at points D or E in figure 8.1. Figure 8.2 magnifies the relevant section of figure 8.1 in order to focus on such shifts. The exogenous action creates an increase in the forest land value between two periods, such that the frontier of natural forest shifts from some previous point E' to a more distant point E in figure 8.2 and creates rents or excess profits equal to the darker-shaded area. This action must be rapid or private investors would anticipate it, invest accordingly, and extract the rents as they gradually arise in small increments. We can imagine two general actions that fit this description: increased rural infrastructure (roads in particular), and effective timber harvest constraints.

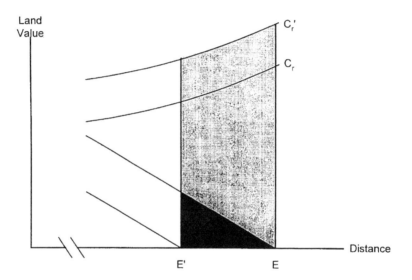

Figure 8.2. Forest rent

Roads are the best example of improved infrastructure that affects forestry. Building a new road into the interior rapidly makes the previously inaccessible forest more accessible. It shifts the intersection of the forest land value gradient from E' to E and creates a rent equal to the darker area in figure 8.2. We often overlook that this rent is a return to roadbuilding, and the authority arranging for the road may have a claim on its collection. For this reason holders of harvest permits or forest concessionaires in many tropical countries design their own roads so they can limit access by unauthorized users. Where they do not restrict access, agricultural settlers and illegal loggers generally follow.[4]

The second potential source of forest rents is an effective forest policy constraint, most commonly a policy restricting timber concessions from select areas or restricting annual harvest levels within a concession. The policy constrains harvests while prices rise (or access costs decrease) over time and the land value gradient moves out to point E in figure 8.2. Once more, a forest rent equal to the darker area arises.

The agency responsible for the constraint (or the road) created the rent, but the existence of this rent raises several questions: Why did the constraint exist in the first place and why does the agency want to relax it and allow logging now? What has changed? If relaxation of the constraint is an appropriate action now, then why was it not appropriate at some earlier time? Are there other forested areas in the country where the policy still constrains harvests? Should the agency relax the constraint in these areas as well? (We could ask similar questions about roads into other new areas.) These are important questions because the constraint forces the economy to forego a real financial gain equal to the rents from all such areas. Any government should be very clear as to why it would make such a decision.

This point is doubly true because effective enforcement of the constraint was probably very costly. We can see this by recalling that the costs of protecting the rights to the forest property c_r exceed the value of forest land at the frontier. (Indeed, the net loss alone is greater than the forest rent that remains because the increasing cost of enforcing property rights means that the lighter area in figure 8.2 is greater than the darker-shaded area.) On the other hand, fully effective enforcement is seldom possible as the government probably could not afford the costs necessary to exclude all trespassers from the forested region between E' and E. An unprotected rent is an incentive for illegal logging, and virtually all forested countries in the world share in a history of illegal logging. A history of illegal logging means that much of the rent has already been captured and the remaining forest and the remaining rent in figure 8.2 must be smaller than anticipated.

The Philippines has two excellent examples of policy constraints that created rents. The first is any adjustment in the annual allowable cut (AAC) that actually leads to an increase in timber harvests.[5] Adjustments in the AAC are government policy and where the previous AAC was a significant constraint, and if illegal logging did not deplete the resource, then a rent accumulated. It awaits collection whenever DENR makes the policy decision to increase the AAC. The second example of rent creation is Lianga Bay. The rents in this case were created by a militant local population (rather than central government authority) who expelled the government-approved logging operation and who still prevent access to the resource by outsiders. Regardless of any political judgment of the local action, it created a rent that awaits collection once loggers are allowed renewed access to the resource.[6]

162   Economics of Forestry and Rural Development

Collecting Forest Rents

Finally, if some effective action created a rent—for whatever reason and at whatever cost—then the next question is the best way to collect that rent. Foresters generally collect rents as product prices, usually stumpage fees or log charges. Figure 8.3 provides a variation on the depiction of forest rent that allows us to focus on log charges. The vertical axis is value per log (rather than value per land unit) and the horizontal axis continues to measure distance or decreasing access—as it did in figures 8.1 and 8.2. The delivered price of forest products is p_f (for example, the price of logs at the mill or at the port) and c_x is the cost of harvesting and delivering the logs. c_x increases with increasing distance from the mill or decreasing access to the forest. Rent, equivalent to the darker-shaded area between E' and E in figure 8.2, is the full area between p_f and c_x. It is the total stumpage value for all standing timber between points E' and E in either figure.

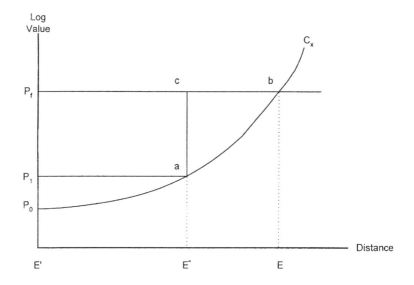

Figure 8.3. A second view of forest rent

Identifying the maximum stumpage value for each log or each increment of standing timber is a practical impossibility. It requires detailed fee appraisals that are different for every logging site. Therefore, many forest ministries rely on a uniform fee per delivered log. (In the Philippines, DENR distinguishes between species groups but the fee remains uniform with respect to the location of log removal or the difficulty of extracting and delivering it.)

In figure 8.3, this is equivalent to collecting p_f-p_1 per log out to point E". It should be clear that, in this case, the rent collected by the government is less than the full rent available (which would be p_fbp_1). Furthermore, uniform fee structures restrict profitable logging to the region between E' and E" and leave an incentive (profits equal to the triangle abc in figure 8.3) to illegally log the region between E" and E.[7] Many policy analyses fail to recognize that avoiding this incentive for illegal logging is generally possible only with enforcement costs that exceed the value of the resource being protected—and this explains why so many countries experience illegal logging.

A long tradition explains forestry's general reliance on auctions of stumpage or logs, but the tradition does not fit the modern forestry situation in most countries. The German and French foresters who were our forebears 100–150 years ago managed the forest and did the logging, and then sold the logs themselves. They needed inventories of standing timber to plan their long-term harvest operations, and they scaled the logs they harvested in order to discriminate by grade in their log sales.

British and North American foresters learned their forestry from these European foresters, and then taught others, including the early faculty at the College of Forestry at Los Banos in the Philippines. Few seem to have re-examined the local context. The European procedures are only partially applicable in North America where public agency foresters often manage the forest but never do the logging. These procedures are not applicable at all in the Philippines and many other places where public agency foresters sell timber harvest permits or timber concessions (an area of forest land, not a log), and a private operator manages the concession and the logging activity.

Forest ministries in general would do well to reconsider their objectives before determining their timber pricing procedures. Any enterprise succeeds best by focusing on the market for its own product. Therefore, the European foresters who did their own logging also established log markets. On the other hand, the Philippine government's foresters do not do their own logging. They will recover the largest revenues by establishing markets for the rights to use land and forests. This means that lump sum auctions for use rights to the specified forest areas would be an improved alternative to log fees—so long as entry to the auction is open, additional bidders are welcome, and the size of the sale is not so large as to limit competition.

Lump sum auctions for the rights to use a specific forest area decrease both government costs and the costs of private loggers, and they insure that the government collects the maximum forest rent. They also modify the government's enforcement activities—allowing it to concentrate on long-term land values consistent with its management responsibilities. This decreases the government costs by removing the need for government forest inventories, government log scalers, and government collectors of log fees. It also removes any reason for the government to monitor the movement of log trucks. In sum, the administrative task of collecting one single lump sum fee is a much simpler task than collecting many individual log fees.

Removing the government scaling and monitoring activities also removes the incentive for loggers to mask the origin of logs from public lands and frees the flows of log trucks. For some loggers, this would save enough log delivery time to allow them to make two loads per day rather than one. This difference decreases costs for the logger, increases the return on the loggers' investments in trucks by 100 percent, and it would enable loggers to increase their lump sum bid prices. Of course, higher bid prices would mean greater government rent collection.

In terms of figure 8.3, private bidders will not bid on land or forests beyond point E because their own costs exceed their potential gains. They will bid up to the full rent for the region from E' to E, but each bidder's judgment of this rent and the best bid itself depends on the bidder's own costs c_x and the price p_f the bidder can obtain for delivered logs. Bidders will rely on appraisals conducted in their own interests. Government timber appraisals are expensive, and they are unnecessary in regions with multiple logging firms and free entry to the auction. The government only needs to mark the sale boundaries, host the auction, and monitor its environmental harvest standards.[8]

Environmental harvest standards and environmental monitoring were unnecessary for the old European foresters because they managed the environment themselves. They could guarantee sustainability or any other environmental standard with their own management decisions. Environmental standards are necessary in the Philippines because the concessionaire wins only a restricted right to use land that will eventually revert to the government. The government and the Philippine public have a continuing interest in the long-term condition of property that concessionaires lease only temporarily. We will discuss the environmental standards in more detail in our recommendations for CBFM and IFMA.

The auctions themselves should be designed to attract multiple bids, and auctions of large tracts which only attract single bids might be divided and rebid with the objective of receiving more bids. Single bid auctions on small sales will be relatively unimportant because they can only occur in locations where the values at stake are so low as to fail to entice competition. Moreover, a small number of bidders for any local timber sale may only mean that only local bidders have the situational knowledge necessary to extract gain from it. In this case, the rent may really be a return on the bidder's local knowledge. Taxing it away would be a disincentive for local entrepreneurial activity. Furthermore, the government's administrative costs for obtaining greater rents on small single bid auctions would rapidly exceed the value of additional rents collected.

Altogether, these insights instruct us to rely on lump sum auctions for land use, not auctions of logs or stumpage and not standardized log or stumpage fees, to focus on locations where the auctions will entice multiple bidders and large bids, and to monitor environmental performance standards—but to minimize government costs and to rely on the advantages of local operations in the many small and general cases of few bids and low bid values. In the Philippines, IFMA fits the former case and CBFM fits the latter.

A Stable Policy Environment

As foresters, we often overlook the importance of a stable economic and political environment. In this case, stability means predictable and consistent expectations about change. (It does not mean "no change.") Unstable macroeconomic and policy environments destroy the incentives for long-term investments like investments in sustainable forestry, and encourage practices like "cut and get out" logging behavior. These can be real problems in this day of ever-increasing environmental demands and changing responses of forestry agencies to these demands.

Both forest plantations and the heavy equipment necessary for large scale logging are long-term investments. In the presence of macroeconomic instability and policy uncertainty, firms will underinvest in plantations and logging equipment because they cannot be sure of obtaining the full return on their investments before the policy environment turns to their disadvantage. Conversely, improving macroeconomic conditions and an increasingly reliable policy environment will induce capital investment. Capital investment in forests is a step toward sustainability. Capital investment in better equipment allows logging firms more efficient use of all inputs, including the land and the trees. This may mean that loggers will harvest farther into the frontier, but it certainly means they will utilize the forest they already harvest more completely. Improved utilization (more lumber recovery per log, greater use of low-valued species, more recovery per acre, and less scrap wood left on the forest floor after harvest) is a frequent environmental policy objective. An improved macroeconomic and policy environment will help achieve it.

An improved macroeconomic and policy environment is an evolving condition. Even when it occurs with the suddenness of revolution, confidence in it and the expectation of its continuation, takes time to build. Therefore, an improved macroeconomic and policy environment is unlikely to create excess profits. Its opposite, however, sudden decomposition of this environment, can destroy long-term capital investments, decrease forest investments, and disrupt long-run forest management. Deacon (1994, 1995) showed that this is a critical source of world deforestation, explaining up to twenty percent of the variation in deforestation across countries.

In terms of figure 8.1, an unstable policy environment means that the forestry enterprise does not have confident expectations about whether government rules defining c_r' will either relax or become more confining. Similarly, the enterprise does not know whether government scaling and log market requirements will relax or become tighter. It may not know whether periodic community or government review of its management plans will force a shift in some well-planned activity of long duration. Finally, it does not know when the government might enforce its rules on illegal logging and void the enterprise's contract—regardless of who actually did the illegal logging. In this unstable policy environment, any firm probably does best to protect itself against the less favorable possibilities. This means the firm would act as if c_r' will be higher and prices will be lower. The first shifts point B' to the left and expands the area of unsustainable open access land management between points B' and B. The second decreases forest land value, thereby delaying the time when sustainable forest investments become financially justifiable.

In terms of figure 8.3, for areas where significant forest rents do exist, an unstable policy environment causes loggers to act as if p_f may fall and c_x may rise. This means there is risk that the efficient harvest level could decline. It induces loggers to harvest toward E as rapidly as they can—before policies do change. Thus, an unstable policy environment delays or decreases sustainable forest practices like plantations, increases the area of unsustainable open access for whatever land use activity, and encourages rapid harvest behavior and lower levels of forest utilization in logging concessions. All are obviously undesirable from the perspectives of both economic development and the environment.

An unstable policy environment is a serious issue, and one that probably has significant implications for Philippine forestry today. We can see this by making two comparisons, one with forestry in Indonesia where many natural timber growing conditions are similar and one with investments in the general Philippine economy. Forest concessions in Indonesia endure and concessionaires do invest in plantations in some places.[9] Neither seems to be true in any substantial sense in the Philippines. Indonesia may provide concessionaires more favorable treatment, and the Philippines may or may not think that this favoritism is justified. That is not the point. Indonesia clearly provides more reliable forest policy expectations. Reliability is a necessary ingredient for long-term investments, and it explains the different levels of forestry success between Indonesia and the Philippines.

The overall Philippine economy in recent years also demonstrates the importance of policy stability for forestry. The EDSA revolution in 1986 improved the political environment but it left investors cautious. They were encouraged, but they watched and waited while they confirmed their favorable expectations. The economy stuttered through a period without much capital inflow. With time, confidence did build because government macroeconomic policies and especially policy toward capital investment remained on a steady path. In time we began to see capital inflows and long-term investment—in private construction, in the consumption of consumer durables, and in public infrastructure. The level of capital investment in the last four or five years has been phenomenal.

Logging equipment and plantations are long-term investments —although neither is expected to last as long as some of the construction we see in Manila or Davao City today. So why is forestry an exception to the recent general Philippine success with long-term capital? One reason may be that the forest policy environment places forestry investors in a very uncertain operating environment. Forestry investors face almost perpetual government review and re-approval process, with serious likelihood of effective challenge to reasonable behavior, as well as numerous impossible standards which could be used at any time to void their contracts with government.[10] The government must find a way to satisfy an environmentally conscious public while improving investment policy expectations if it wishes to induce serious long-term private forest management. And it must persevere in this objective because, as we saw following the EDSA revolution, investor confidence does not build overnight.

The positive side of this message is that the Philippines competes in a world of unreliable suppliers for tropical hardwoods and tropical hardwood products. Many of the other producer countries also suffer uncertain macroeconomic and policy environments. Improving the policy environment should give the Philippines a comparative advantage over these competitors, transferring some long-term forestry investments from those countries to the Philippines, and further improving the prospects for sustainable plantation management and increasing government forestry revenues in the Philippines.

Two Applications of the Principles: Opportunities for Successful Community-Based and Industrial Forest Management

More than 200 Timber Licensing Agreements (TLA) covered 2/3 of the forested area of the Philippines in 1976. Some have lapsed and some have been withdrawn. In their place 2.5 to 3.0 million hectares of forest remain, and the deforestation and unsustainable practices that caused all but 28 TLAs to lapse or be withdrawn continue despite the discontinuance of the timber concessions. Of course, an upland population of 10 million is largely responsible, and a substantial flow of migrants regularly increased this upland population and reinforced local deforestation and unsustainable land use through 1997.

Centralized forest management is a daunting task. Its difficulties have induced a global search for alternatives, and the preferred alternatives generally

168   Economics of Forestry and Rural Development

include some degree of a) local community management, particularly for the many cases of lower-value forest resources and b) private sector management, particularly for the smaller number of high-value and narrower purpose plantations and production forests. DENR seems to support both concepts. In the first case, the communities with the greatest stake in the resource would manage it for their own long-term interest—but with oversight from DENR. In the second case, DENR divides private sector management into two classes, but the general rule is to transfer land use to the private firm offering the best guarantee of sustainable forestry. In all cases, the community organization or the private firm agrees to pay a fee to the government for its forest products, generally its logs. The problem is to identify the best means of transferring responsibility from DENR to the local communities or the private firm, while insuring equitable distributions, sustainable practices, and financial returns sufficient to repay foreign loans for the forestry sector—and while also improving the agency's public credibility. This is no easy task—but it is a task common to many forest ministries.

This final section of our chapter recommends simple structures for accomplishing this task, first with CBFM and then with IFMA, and it concludes with a recommendation of new roles for DENR, including the advantages of a stronger environmental presence.[11] As we consider our recommendations, we must remember the principles we just reviewed. We must also recognize that the best program design is simple—because complex designs are costly to administer, because forests are generally low-value resources, and because we know from experience that many of DENR's prior management strictures were difficult to enforce at best. Furthermore, many of the old forestry rules and procedures are poorly designed or even unnecessary for DENR's objectives. They never did fit Philippine forestry administration.

General Recommendation

Our first recommendation is to clearly distinguish between CBFM and IFMA. These are parallel programs designed to address different problems and to satisfy different objectives. Each has its own independent purpose.

CBFM is intended to assist local communities. It is central to our rural community development interests in this book. CBFM reflects an underlying distributive objective of DENR and some of its critics. Locally specific knowledge gives additional value to the areas of CBFMs, and local use made prior DENR management of these areas difficult to enforce.

IFMA is better-designed for those larger and more valuable areas where large-scale private operations can successfully invest and obtain a commercial return on their investments. The important characteristics of a successful IFMA are the opportunity for a fair return on private investments of financial and material capital, significant opportunity for the collection of government timber revenues, and some reasonable opportunity for sustainable management.

Recommendations for Community-Based Forest Management

The basic problems in establishing a CBFM agreement with a local community or community organization are the selections of land use boundaries and of the local group to assign the land use rights. The current Philippine plan for public meetings can help but the public meetings should focus on two simple objectives: refining DENR's preliminary identification of boundaries and identifying interested individuals and local organizations with the capability to manage a CBFM. DENR does not need to face the plethora of issues that can arise in community debate. It does not want to be held accountable at a later date for some prior and obscure discussion. It should announce its intention to contain the discussion of these public meetings within these two topics, and then it should manage these meetings accordingly.

The second step is to draw final boundaries and to remove the most environmentally sensitive areas from within those boundaries, probably the steepest slopes, the stream banks, and specific that sites are unusually sensitive to biodiversity objectives. The Philippine constitution requires DENR to remain the land owner and, like any other land owner, DENR is desirous of receiving its land in good condition at the end of the CBFM lease. DENR also has a responsibility to ensure that the uses of its land neither damage long-term social values like biodiversity nor damage adjacent lands and their uses. The best way to insure these conditions may be to withdraw those environmentally sensitive areas that are within the CBFM boundaries and to maintain them as a direct DENR responsibility. Direct DENR responsibility for the environmentally sensitive areas is simpler and less intrusive than trying to enforce DENR's own management insights, insights which are certain to change over the lifetime of a 25-year renewable lease.[12]

The current CBFM documents require an inventory of the standing timber volume, a formal management plan, and the assistance of a qualified professional assisting organization (AO) to help prepare them. These are unnecessary if DENR has ensured its environmental responsibility. The inventory of standing timber may not even refer to the non-timber resources that give the forest greatest value for the local community. These activities may be beyond local capability and comprehension, and the notion of an AO to advise on them is a denial of the CBFM intention to rely on local judgment. Furthermore, the inventory, the plan, and the AO are expensive. By definition, CBFMs are areas of generally lower-value which may not support these costs. They are also areas where DENR has had limited prior success in enforcing outside rules—like harvest levels based on a timber inventory and a management plan. DENR's ability to enforce its management criteria will not change with the development of a plan or the contractual arrangements of a CBFM. It should eliminate these requirements from its CBFM contracts.

The third step is to transfer the use rights for the land within the final CBFM boundaries. Any local individual, collection of individuals, or organization that can sign a legal contract and which has the capability to manage a CBFM should be an eligible agent. DENR has no reason to discriminate in favor of, for example, licensed cooperatives, or any other bidder group. It has no reason to discriminate in favor of particular land uses (such as timber production). If interest is weak, then DENR might donate the entire parcel of land to the community group or groups which it judges as most responsible. If local interest is strong and multiple local groups display an interest, then DENR might divide the CBFM among the groups—or conduct an auction.

Subsequent to transferring the rights to land use, DENR's continuing responsibility is to ensure a good climate for sustainable land management. The most encouraging situation would be described as "no continuing management oversight, no annual plan preparations, no annual review with its own uncertainty and potentially delayed outcomes, and no restrictions on the marketing of forest products." Permission for land transfers from the successful first recipient of a CBFM contract to new groups after the contract has been awarded will ensure higher valued land uses, higher auction prices, and the efficient scale of local operations, and it would permit the adjustments to each of these features that must occur over the 25-year life of any contract. The likelihood of sustainable land use activities will increase as a result. If DENR's objective is sustainable land use, then the only terms of the original CBFM contract should be environmental protection. DENR has a responsibility to monitor the protected areas, and damage to the adjacent and withdrawn areas previously identified as environmentally risky would be the only grounds for voiding a contract.

In addition, DENR could assist with its objective of sustainable management by providing the assistance of extension-like forestry advice. If this advice is provided with the intention of satisfying the land user's objectives, not the forester's notion of "good forestry," then technical assistance will help build local good will and insure better protection of those environmentally critical areas which DENR withholds for its own management.

The basic CBFM documents require a Trust Fund composed of thirty percent of gross receipts from sales of forest products. The Trust Fund would be used to reimburse DENR costs and the costs of an AO, and also to provide a source of revenue for approved local investment. We believe that the Trust Fund constrains local management and is unlikely to recover many DENR costs. Together, these two fees amount to 44 percent of gross receipts, a huge fee for the most profitable venture, and a totally unlikely fee for local CBFM operations in largely subsistence communities.

In terms of figure 8.4, any management restrictions, including restrictions on use of the Trust Fund, raise local costs to c_x' or reduce market opportunities to p_f' in a location which we already know to be financially marginal. Extracting 44 percent of the sale value of forest products is almost sure to remove all remaining financial incentive, therefore all incentive for sustainable management and all opportunity for DENR cost recovery, from many remaining CBFM opportunities.

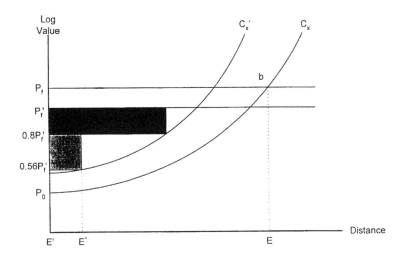

Figure 8.4. The CBFM Trust Fund — and other potential forest revenues

Figure 8.4 describes the case of those CBFMs that might still retain some financial incentive after removal of a thirty percent fee for the Trust Fund. Costs are higher by the expenses for hiring an AO, completing an inventory, and a management plan, etc. Prices are also lower due to government constraints on acceptable activities, including constraints on log shipment. The basic twenty percent fee ($p_f' - 0.8p_f'$) is removed (the larger shaded area) and a second fee equal to thirty percent of remaining receipts ($0.8p_f' - 0.56p_f'$) is removed (the smaller shaded area) to create the Trust Fund. This Trust Fund must cover all previous DENR costs, plus an additional charge for scaling logs and collecting the log fee. The local holders of CBFM rights now have an incentive to illegally log the entire area between E" and E. Obviously, this is not a desirable outcome for the holders of CBFM rights or for DENR cost recovery, it explains why few communities have desired to enter into CBFM agreements, and it explains our recommendation to minimize these fees and to rely on a single basic charge for land use rather than log fees and a Trust Fund.

In sum, our recommendation is to keep both the CBFM contract and its oversight as simple as possible, to minimize DENR costs, and to leave as many incentives for sustainable local management as possible—but to withdraw and protect the most risky environments.

Recommendations for Industrial Forest Management Agreements

The government's objectives for IFMA are a) the collection of forest rents—or the government's constitutional "production share"—and b) environmental protection or sustainable management. There is no equity objective within IFMA. The lands within an IFMA may include degraded forests, forest plantations, and productive residual forests but formal distinctions among these land use classes are unnecessary because they do not help satisfy these basic IFMA objectives. They only raise DENR's costs for assessing the differences and create incentives for private operators to misclassify their lands.

Once more, the basic problem is to identify the IFMA boundaries and the environmental risks within these boundaries. Once more, there is no need for the standard forest inventory, but an assessment of environmental risks would be useful. Public participation can assist in identifying boundary conflicts and environmental risks and the most helpful formal public participation activity would focus on these two issues.

DENR may choose to withdraw the areas of environmental risk from the IFMAs or it may choose to leave these areas in the IFMAs while establishing standards for their management and monitoring performance. The environmental bond referred to in the current IFMA documents was designed to protect against environmental non-compliance. For the bond to satisfy this objective, it must be set at a level that would cover DENR's costs of rehabilitating the environment in the event of non-compliance. Therefore, it will be a function of the erosion control or reseeding or other activities specific to the particular IFMA and necessary to restore that environment if the private operation defaults on its contractual obligations. Each IFMA bond would be different, and DENR would retain the bond only in the event of default on the environmental performance with respect to contractually identified environmental risks.

The government does not receive its constitutional "production share" from the bond. The government's production share comes from payment for the use rights for land within the IFMA boundaries (and subject to management constraints for the areas of environmental risk). The payment need not

Secure Forest Tenure, Community Management, and Deforestation 173

distinguish between degraded, plantation, or productive residual forest because private bidders will make their own assessments of the values of each, and they will bid more where the land or standing residual forest are more valuable. Thus, we are recommending both an environmental bond and also an auction for the land use rights.[13] Their values are unrelated and each is unique to any particular IFMA. The bond would not be retained except in event of environmental non-compliance. The value of the winning bid at the auction is the government's production share and it would always be collected.

DENR may choose to set a the land use fee, or it may choose to set a minimum bid price at which an auction begins and it may choose to withdraw any offers for which there are an insufficient number of bidders. In the latter event, DENR must ask itself whether the sparse number of bids was because the particular IFMA is financially unattractive at the minimum bid price, or whether redrawing the boundaries (perhaps dividing the IFMA into multiple smaller IFMAs) would attract more bids.

In any event, the optimal IFMA size from the perspective of DENR has to do with balancing the maximum auction value against DENR's continuing cost of environmental monitoring. The optimal size for private management will change as time passes and forest values change. DENR will assure itself of the highest-value land uses and the greatest receipts from the auction if it permits IFMA transfers, in whole or in part, to any operator willing and capable of honoring the initial contract, including the bond on environmental performance.

The large backlog of areas currently eligible for IFMA consideration might lead DENR to consider the sequence of its auctions. All IFMA auctions cannot be held immediately because the boundaries and environmental standards for some have yet to be established. Those potential IFMAs which might be most responsive to environmental improvement might be bid first. Those areas where the presence of large and active private harvest operations might exercise useful controls on illegal local uses might also be bid early. Those IFMAs with internal areas of large environmental risk and those where active harvest operations would provide inadvertent access to areas of adjacent environmental risk might be delayed. For example, DENR might delay scheduling bids on an IFMA that is adjacent to an old growth natural forest preserve and which would provide the only access to that forest. The delay would ensure a period of greater protection for the biological reserve.

Subsequent to awarding the contract, DENR's only responsibility, albeit a critical responsibility, is to monitor environmental performance on the risky sites within the IFMA according to the standards for those sites laid out in the contract. A good private operation will create a management plan for its own purposes but DENR approval would be necessary only for that part of the plan addressing the sequencing of environmental improvements like contractual reforestation requirements. DENR need not impose an annual allowable cut (AAC) restriction because we know it cannot monitor such restrictions successfully, and because the reason for the AAC restriction is to protect the environment. Specific environmental objectives will be accomplished with other contractual obligations and DENR's own more focused environmental monitoring. There would be no reason for periodic reviews of any IFMA beyond the regular environmental monitoring and the review DENR would initiate itself upon evidence of environmental non-compliance.

The result of our package of recommendations should be a simpler process for granting IFMAs and a more focused process for monitoring their performance. This means lower DENR costs, lower costs for the private operation, greater investor incentive and greater confidence in the long-run reliability of the process, and an improved likelihood of long-term and sustainable forestry investments. The financial expectations of the private operator will be greater for all of these reasons, and the government fee recovery will also greater. An improved climate for private operations is important. Many of the 227 IFMAs approved since 1992 have not been implemented because they could not obtain financing. Meanwhile, DENR's responsibilities are also simplified to emphasize environmental protection and this too should improve accomplishment of that fundamental objective.

Conclusion: A New Role for DENR and Its Foresters

Our recommendations all point to a reinforced role for DENR foresters as protectors of the natural environment. DENR's role in the full range of forest management and especially timber management activities would decrease, but its revenue recovery would increase and its role as an environmental monitor and enforcer would expand. The public is a severe critic of many forestry agencies around the world, but we anticipate that the Philippine public would respond well to this idea. The private operators of CBFMs and IFMAs will also respond well so long as DENR is consistent in its environmental decisions.

Environmental risks are pervasive, but DENR's resources are limited. Therefore, DENR should focus its environmental monitoring and enforcement resources on areas that satisfy the combined criteria of a) great risk and b) potential DENR enforcement. One without the other is not very meaningful. DENR managers will do well to remember this rule when removing risky sites from CBFMs and selecting environmental standards for risky areas within the IFMAs.

We have shown that many forests and degraded natural environments will remain open access areas subject to unsustainable land uses (region B' to E in our figure 8.1), and that DENR is unlikely to alter this fact without bringing an unacceptable level of resources to bear on the problem (c_r' in figure 8.1 is too high). DENR's best option is to select the most threatened and riskiest environments within this range of open access, and also the most important environmental resources in the neighborhoods of current logging and other forest removal (points D or E in figure 8.1). It might choose to maintain a list of those environmentally critical locations that is commensurate with its abilities to monitor and enforce. The list will change with time but the act of thinking about the choices on the list may gain the public's good will and alert the worst potential violators.

DENR can shift some of its field personnel away from responsibilities for forest inventories, timber management plans, log scaling, and log fee collection. These personnel can assist with the agency's expanded environmental monitoring and enforcement responsibilities, and they can gain local good will for the agency by providing extension-like forestry advice in select areas where agroforestry has potential (point F in figure 8.1).

Central office personnel can participate as well. They can search for ways to decrease the costs of transferring and administering the land use rights (c_r' in figure 8.1) to CBFMs and IFMAs, and thereby decrease one part (B'–B) of the area of unsustainable open access. The policy office in DENR can participate by identifying macroeconomic policies and policies that favor other sectors of the economy but spillover to affect forestry or to send migrants to the uplands and, in either case, destroy forest environments. DENR has an important stake in the economy-wide policies that affect forestry. This stake is not commonly recognized by the European and North American forest ministries, but it is important for many developing countries where these policy spillovers can have critical impacts on the forest environment. DENR's Secretary and other policy makers outside of DENR need to understand these issues better in order to successfully protect the upland environment.

This chapter was prepared for the specialized current policy interests of Philippine forestry. Nevertheless, the general problems that the Philippine's faces—sustainable land use, government recovery of revenues, and conflicts between public land managers and local community interests—are common to forestry almost anywhere in the world. The general interest in the social forestry topic of this book is one recognition of the importance of local solutions to these problems. Our Philippine insights, especially for CBFM, will be applicable for the design of public agency social forestry programs wherever they occur.

NOTES

1. NRMP, Manila; NRMP; NRMP; and Centre for International Forestry Research, respectively. NRMP is a Philippine government-USAID project examining

forest land use policy and management. M. delos Angeles, A. Juhola, and Juan Seve provided critical comment.

2. Tomich et al. (1995, chs. 4,5) explain that the costs of protecting property rights increase with the size of the property—as well as with distance. DENR manages the largest unit of land in the Philippines. Therefore, its costs for protecting the property are probably greater per hectare than the costs of any other landowner in the country. This is a related, but additional, reason to transfer some DENR properties to upland settlers or to lower-cost community forestry operations.

3. For example, Hofstad (1995) describes rising charcoal prices and ever more-distant wood removal in the region surrounding Dar es Salaam, Tanzania; Chomitz and Gray (1994) describe a similar pattern for roads, agriculture, and deforestation in Belize; and Krutilla et al. (1995) provide a more general example for deforestation around 33 urban centers in Africa, Asia, and Latin America. Alston et al. (1996) describe the relationships between settlement, property rights, and rising prices on the frontier of Brazil's Amazon. Both Johnson and Libecap (1980) and Berck (1979) show that 19th century US logging never extended into the frontier at more than a financially optimal rate.

Several recent analyses describe situations where prices have risen sufficiently to induce investments in sustainable forestry. See Hyde and Seve (1993) for Malawi where subsistence household plantations may overcome a 3.5 percent rate of forest depletion within ten years; Scherr (1995) and Patel et al. (1995) for western Kenya which shifted from ninety percent dependent on natural forests for construction wood in the 1950s to eighty percent dependent on forest plantations today; or Amacher et al. (1993b) and Singh (1994) for similar stories for Pakistan's Northwest Frontier Province and India's Punjab, respectively. Amacher et al. (1992, 1993a) examine two districts in Nepal and describe the production, price, and consumption differences that cause subsistence farmers to plant trees and adopt wood-saving technologies in one district but not the other. Vianna et al. (1995) show that, once the market price rose high enough, even Brazil nuts, for many the forest product that has been least responsive to scientific intervention, have been domesticated in commercial plantations.

4. Both Rice and Gullison (1995) and Robbins et al. (1995) describe concessionaires who control road access into Bolivia's Amazon in order to protect their own investment. Chomitz and Gray (1996) describe how agricultural development and deforestation in Belize follows road development. Thailand's military used this principle to their advantage. They built roads into Thailand's forested Northeast and induced settlement in that region in order to create a human buffer to aggression from neighboring states. Similarly, the Lao government offered harvest concessions adjacent to the new Bukeo-Boten road into China as partial payment for the road's construction.

5. The Philippines, like most countries, calculates its AAC from a biological formula that encourages underharvesting in some regions and permits overharvesting in others. See Hyde (1981) for a review.

6. The Philippine log export ban could be a third example. An effective log export ban would cause a rent to build up—just as an effective AAC does. In general, enforcement is difficult (c_r' is high), illegal logging tends to occur, the rents dissipate, and log export bans are not especially effective.

7. The maximum stumpage fee p_r-p_o will not maximize rent collection because it minimizes the number of logs. The maximum rent collection possible with

a uniform fee is a mathematical function of c_x (or a function of the elasticity of extraction and delivery costs). This maximum is an empirical question that varies with each logging site and each timber concession. Clearly, a broad regional or national logging fee does not satisfy any obvious objective, imposes significant collection costs, and invites illegal logging. Just as clearly, raising the uniform fee will not necessarily increase the rent collection. Indeed, Indonesia has raised its fee several times in recent years only to find that the government's total collection of forest revenues declines.

8. The government cost savings are not trivial. In the United States in one recent year the costs of government timber sales exceed timber sale receipts on 62 of 156 national forests. Boado (1988) claims that Philippine timber operations, like government timber operations in the US, cost more than they return to the Philippine treasury. Surely, the expensive costs of inventories, appraisals, scaling, and stumpage fee collection and enforcement are part of the problem. All of these costs are unnecessary with lump sum auctions for use of a land area, rather than auctions of logs or stumpage.

9. Or they did before Indonesia's financial crisis of 1998.

10. For example, TLA renewals were always in doubt, and the required procedures for allotting IFMAs had already changed three times by 1997. Extensive and regular NGO, local government, and DENR review written into current CBFM and IFMA policy extends the uncertainty.

11. The basic documents for CBFM policy are Executive Order 263 and DENR Administrative Order 22–1993. DENR memorandum 95–18 which sets up eleven pilot sites, and Community Forest Management Agreement 13–001 with the SAMMILIA Forest Development Cooperative (Lianga, Surigao del Sur) are also useful references. The basic documents for IFMA are DENR Administrative Order 60–1993 and the IFMA Implementation Manual.

12. For example, some CBFM documents ban mechanical logging. This ban unnecessarily constrains CBFM activities, and it constrains the opportunity for financial gain in non-erosive areas. Furthermore, it will become less necessary as the world designs better logging equipment. It would be unnecessary now if the erosive areas subject to damage by mechanical logging were removed from the CBFM base.

13. Possessors of the 28 existing TLAs have a right to request the transfer of their old TLAs to IFMAs. This transfer can be accommodated within our recommendations by allowing the possessor of the old TLA the right of first refusal at the established fee or, in the event of an auction, at the auction's high bid price.

REFERENCES

Alston, L., G, Libecap, and E. Mueller. 1996. *Emerging land rights in the Amazon.* Champaign, IL: Draft manuscript, Economics Department, University of Illinois.

Amacher, G. S., W. F. Hyde, and B. R. Joshee. 1992. "The adoption of consumption technologies under uncertainty: the case of improved stoves in Nepal." *Journal of Economic Development* 17(2):93–105.

Amacher, G. S., W. F. Hyde, and B. R. Joshee. 1993a. "Joint production and consumption in traditional households: fuelwood and agricultural residues in two districts of Nepal." *Journal of Development Studies* 30(1):206–25.

Amacher, G. S., W. F. Hyde, and M. Rafiq. 1993b. "Local adoption of new forestry technologies: with an example from Pakistan's Northwest Frontier Province." *World Development* 21(3):445–54.

Berck, P. 1979. "The economics of timber: a renewable resource in the long run." *Bell Journal of Economics* 10(2):447–62.

Boado, E. L. 1988. "Incentive policies and forest use in the Philippines." In R. Repetto and M. Gillis, eds. *Public Policy and the Misuse of Forest Resources*. Cambridge: Cambridge University Press.

Chomitz, K. M., and D. A. Gray. 1994. "Roads, land, markets and deforestation: a spatial model of land use in Belize." Washington: World Bank Policy Research Working Paper 1444.

Deacon, R. T. 1994. "Deforestation and the rule of law in a cross section of countries." *Land Economics* 70(4):414–30.

Deacon, R. T. 1995. "Assessing the relationship between government policy and deforestation." *Journal of Environmental Economics and Management* 28(1):1–18.

Department of Environment and Natural Resources. no date. *1994 Philippine Forestry Statistics*. Manila: DEMR, Forest Management Bureau.

Hofstad, O. 1995. "Woodland deforestation by charcoal supply to Dar es Salaam." As, Norway: Agricultural University of Norway draft manuscript.

Hyde, W. F. 1981. *Timber Supply, Land Allocation, and Economic Efficiency*. Baltimore: Johns Hopkins University Press for Resources for the Future.

Hyde, W. F., and J. Seve. 1993. "The economic role of wood products in tropical deforestation: the severe example of Malawi." *Forest Ecology and Management* 57(2):283–300.

Johnson, R. N., and G. D. Libecap. 1980. "Efficient markets and Great Lakes timber: a conservation issue reexamined." *Explorations of Economic History* 17:372–85.

Krutilla, K., W. F. Hyde, and D. Barnes. 1995. "Urban energy consumption and periurban deforestation." *Forest Ecology and Management* 74(2):181–95.

Patel, H. S., T. C. Pinckney, and W. K. Jaeger. 1995. "Smallholder wood production and population pressure in East Africa: evidence of an environmental Kuznets curve?" *Land Economics* 71(4):516–30.

Scherr, S. J. 1995. "Economic factors in farmer adoption of agroforestry: patterns observed in western Kenya." *World Development* 23(5):787–804.

Singh, H. 1994. "Reforestation in the Punjab—a consequence of the Green Revolution." Cambridge: unpublished manuscript from Harvard University JFK School of Government.

Tomich, T., P. Kilby, and B. F. Johnson. 1995. *Transforming Traditional Economies*. Ithaca: Cornell University Press.

Viana, Virgilio M., Ricardo A. Mello, Luiz M. deMorais, and Nilson T. Mendes. 1996. "Ecology and management of Brazil nut populations in extractive reserves in Xapuri, Acre." Manuscript submitted to Bioscience.

Yin, Runsheng, and William F. Hyde. 1996. "Trees as an agriculture sustaining resource: evidence from northern China." *Agroforestry Systems* (forthcoming).

CHAPTER 9

Rural Reform and China's Forestry Sector: Rational Farmer Expectations and the Case for a Stable Policy Environment

Runsheng Yin, David H. Newman, and William F. Hyde[1]

China's rural reforms, beginning in 1978, enabled remarkable growth. The agricultural sector grew at a rate of 7.7 percent annually until 1984, and it has continued growing at more than four percent annually ever since (China Agricultural Yearbook 1990–92). The forestry sector benefitted from similar reforms and it includes many of the same farm households, but growth in this sector has not been uniform. The experience of the northern forest production region parallels the general agricultural experience as both annual timber harvests and investments in forestry have increased dramatically. In the southern forest production region, however, the existing forests have been drawn down without replacement. Forest cover more than doubled in the north, while it declined by more than ten percent in the south.

What explains the different experiences of the two regions? We believe the answers can be found in sharp regional differences in the implementation of forest policy reforms. Both regions experienced increases in the share of private tenure for their forest lands and both experienced improving market incentives, but government authorities were slower to liberalize and quick to recant on some reforms in the south (Wang et al. 1991, Yong 1987). The resulting uncertain policy environment explains the southern region's poorer performance.

These contrasting regional experiences demonstrate the critical importance of investor expectations and stable policy environments for long-term investments like forestry. Favorable policies at any moment in time are insufficient. Favorable investment decisions also require a predictable longer-term policy environment. Otherwise, we can anticipate that investors will take short-term advantage of any new incentive, and disregard the less certain longer-term opportunities. In forestry, favorable incentives and a stable policy environment will encourage current harvests and also the investments in the standing timber that will grow into future harvests. Forest landowners and managers will harvest without reinvestment, however, if they are uncertain that the new incentives will be lasting. Landowners will increase their current harvests beyond any notion of the long-term sustainable harvest level in order to capture the short-term opportunities while their personal claims on the timber are still valid.

Policy makers generally do understand the importance of stable policy environments, but they often overlook the importance of broad-based stability when examining selective natural resource sectors. This is a crucial oversight for countries experiencing rapid economic transition and for sectors like forestry where the basic investments are all long term. Forest plantations take decades to develop and even harvests from mature timber stands require investments in roads, logging equipment, and sawmills that may endure for decades. In such cases, sharp and fluctuating policy modifications can eliminate the incentive to invest in either the forests, the roads and logging equipment, or the processing facilities.

China and the countries of Eastern Europe are the common examples of economies in critical transition. Nevertheless, the same questions of economic transition and stable policy environments are also important for countries like the Philippines that are emerging from extended periods of political and economic repression, or countries like Bolivia that are contemplating significant new forest development and the new policies necessary to manage it. Deacon (1994) provided one measure of the overall impact of policy instability in forestry. He estimated that political instability explains twenty percent of the variation in national rates of deforestation. Our estimates of regional forest supply in chapter 6 support Deacon's estimate.

China's experience with rural forestry reforms offers an unusual opportunity to examine this idea more closely. China's experiences with similar reforms in two different regions can be used to demonstrate the sources of policy improvement, and also the contrast between stable and unstable policy environments, on the many small investors characteristic of farm forestry. We will review this experience, first discussing the general patterns of China's rural forestry reforms, and then developing a supply-response model that accounts for differences in short-term timber harvests and longer-term forest investment behavior. Estimates of this model for the two regions in China and the period since 1978 will show the impacts of farmer expectations about forest policy. Evidence from surveys of household attitudes toward government policy will support our model and confirm our view that unstable policy environments cause investors to maximize short-run gains and disregard long-term price increases and improvements in land use rights that might otherwise favor investments in sustainable forest management.

China's Rural Reforms and Farm Forestry Behavior

Following the collective movement of the late 1950s, China's rural sector was organized under the production team system. Workers received payments based on the average product of the collective as a whole rather than on their individual marginal products. As a result, incentives to work were low, and productivity and living standards stagnated (Perkins and Yusef 1984).

After the Cultural Revolution, China began to shift its political course in the direction of liberalization. Farmers recognized the signs of change and reacted by creating alternatives to the old system. These alternatives soon converged as the Household Responsibility System (HRS). Under HRS, collectively owned farmland and other assets were distributed to individual households, and households rather than production teams became the basic production unit.[2] Each household became responsible for its contractually assigned output quota, and for its share of taxes and the collective's common expenses. Households retained all production in excess of their contracted production quotas and they now received their own marginal products for outputs in excess of these quotas. The collective's common costs, and household contributions to them, declined because external supervision and evaluation by the collective was no longer necessary (Lin 1992).

Several assessments of China's rural reforms (Lin 1992, McMillan et al. 1989, Wen 1993) all suggest that the transition from the collective system to HRS was the primary source of agricultural growth in the late 1970s and early 1980s. The improved incentive structure under HRS would have been of limited relevance, however, without expanding market opportunities because even subsistence households produce some goods for market exchange, and all households produce for market exchange once their production levels exceed the level necessary for household subsistence. The government's compulsory delivered prices, and even its above-quota prices, were always lower than the open market prices (Li et al. 1988, World Bank 1985). Therefore, the development of well-functioning markets was a second important production incentive.

Farm Forest Policy

China's forests are classified by their regional and management characteristics. The Northeast/Southwest National Forest Region (NSNFR) is the China's major timber supply region. Its forests are centrally managed. The Southern Collective Forest Region (SCFR) is the second most productive forest region. The Central/North/Northwest Farm Forest Region (CNNFFR) covers the remainder of the country. The western part of the CNNFFR is a dry area that is adverse to forestry, but the eastern part is a major agricultural area. We will show, in chapter 11, that vigorous agroforestry activities since the late 1970s have significantly altered the landscape in this eastern part. The government's rural policy reforms have affected farm forestry in the SCFR and in the eastern CNNFFR (which we will call the southern and northern forest production regions, respectively), but they do not affect natural forests in the NSNFR or in the more arid western CNNFFR.[3]

The Regional Experiences

As HRS swept China's agricultural sector, questions regarding its validity for forestry arose as well. The authorities in the southern forest production region were not initially responsive. They denied household rights to forest land on the presumption that household operations were too small to support longer-term forest development activities. Two conflicts arose, one due to the denial of property rights and one due to boundary conflicts with entrenched state forests on collective lands. In general, only fragmented and less productive forests and woodlots were contracted to household management and, by the end of 1984, more than seventy percent of the region's forests still remained under collective control. Typical households only managed about one hectare of land for their own use (Xin 1985).

Eventually, popular objections overcame the government reluctance and the southern authorities slowly responded by adopting a contract system like HRS for collective forests.[4] Contracted timberlands and plots of private forest were merged until households held contracts with the collectives for more than sixty percent of all cooperative forest land by 1986. Contractual agreements extended up to fifty years. Quotas became a popular contract criterion and households were entitled to retain their production surpluses.

The government in this southern production region also opened the timber markets in 1985. Prices rose rapidly and induced further harvests and speculation. Nevertheless, southern farmers failed to expand their reforestation and forest management activities. They were not convinced that the government would sustain its policies (Wang et al. 1991, Yin and Xu 1987).

Unfortunately, the farmers' fears were justified. The government became concerned about the liquidation of forests and the stagnation of forest management. It reversed its policies in 1986. First, it returned timber markets to the control of state procurement companies and restored tough regulations on harvest volumes. For those trees that remained under household responsibility, farmers had to apply for harvest permits and they could sell only to the state procurement companies at prices that were approximately half of the market prices. Subsequently, the government began re-consolidating contracted timber lands. It also raised forestry taxes and fees on farmers in order to fund the government's own reforestation activities.

The experiences of the northern forest production region were very different. The government paid forestry less attention in this region because timber production was less important. It was easier for forest policy reform to follow the pattern of the agricultural policies that are relatively more important to this region. A household forest tenure system called "land carrying trees" quickly followed HRS for agriculture. Trees on or near contracted agricultural land were assigned to agricultural households and bare lands suitable for trees were also re-allocated to households. A few larger shelterbelt networks and commercial forest plantations that remained in collective ownership were also contracted to household management. Overall, the share of contracted forest land in the northern forest production region rose to 91 percent by 1984 (Yin 1994).

Furthermore, the government allowed farmers in this northern forest production region the freedom to sell their timber at market prices. Harvest permits were unnecessary, sales to state procurement companies were not required, and taxes were low—ranging from three to eight percent of sales revenues. In general, these arrangements were acceptable to farmers, and they have prevailed (Zhong et al. 1991, Yong 1987).

The Numerical Evidence

Tables 9.1 and 9.2 summarize the experiences of the two forest production regions. Table 9.1 summarizes the price and household landholding experiences between 1977 and 1990. The price evidence is for two sample provinces in each region. Real timber prices rose over four-fold in all cases, with larger absolute increases in the northern production region. These price increases should have been an incentive for forest investment. The southern policy shift from procurement prices to retail prices in 1985 doubled that region's price incentive but even retail prices in that region were controlled at a level well below the market prices. Of course, the south's subsequent return to procurement prices removed the additional price incentive and confirmed the existing policy uncertainty.

HRS represents an increasing opportunity for household decision making. As a measure of HRS in table 9.1, we use the share of agricultural and forest land previously managed by collectives that was converted to household contract. This measure is better than a measure of purely forest land because some farmers chose to reallocate land between agricultural and tree crops.[5] The household shares rose more rapidly in the north after reforms beginning in 1980. They were ninety percent complete by 1984. Household shares rose more slowly in the south, and the government even rescinded some previous household contracts. Thirty-five percent of forests on southern collectives were still excluded from household management in 1990.

Table 9.1. Regional Comparisons of Increasing Timber Prices and Improving Tenure

Year	Northern production region				Southern production region					
	Fuyiang	Suxian			Fuzhou		Yichun			
	Market price	Market price	HRS share		Procurement price	Retail price	Procurement price	Retail price	HRS share	
1977	119.8	117.6	0.0		28.3	66.1	23.0	66.5	0.0	
1978	125.4	120.7	0.0		25.0	66.6	25.3	67.7	0.0	
1979	136.3	122.0	0.0		38.1	70.1	35.3	73.4	0.0	
1980	165.9	152.0	32.9		46.3	83.3	50.5	87.8	14.4	
1981	201.3	184.0	67.0		45.8	97.1	61.1	97.7	29.5	
1982	241.0	222.0	78.8		56.5	104.3	64.0	108.1	37.7	
1983	262.2	255.1	84.5		48.8	105.5	55.0	109.7	43.3	
1984	265.0	297.0	91.0		55.3	109.0	57.9	116.4	52.4	
1985	310.8	368.9	91.0		84.0	174.5	99.5	165.9	64.8	
1986	351.5	377.3	91.0		93.0	207.5	118.0	226.2	64.8	
1987	436.0	457.3	91.0		126.0	289.8	177.4	346.3	64.8	
1988	630.1	667.8	91.0		171.4	444.1	173.1	502.2	64.8	
1989	741.7	787.0	91.0		169.3	436.7	189.9	473.2	64.8	
1990	639.0	698.0	91.0		170.4	350.3	164.8	386.4	64.8	

Notes: Prices are nominal, averaged across species and grades. Price data are from financial balance sheets of local timber companies and government reports. HRS is the ratio of forest land allocated to households to total collective forest land. HRS shares are from the provincial forest services.

The rate and final level of conversion to HRS distinguished the southern experience from the northern, but the critical factor was uncertainty. The southern authorities reluctance to complete the transition to HRS, and their recision of some HRS contracts, coupled with their decision to return to procurement prices, left southern farmers in a state of uncertainty with respect to future prospects for forestry.[6]

What were the farmers' responses to policy reform? Table 9.2 shows the levels of timber harvests and forest cover for two new sample provinces in each region for two critical years: 1977 and 1987. The 1977 data provide a baseline from which to assess policy departures. The 1987 data reflect a time period sufficient to absorb the re-adjustments in the southern production region that occurred after authorities rescinded the earlier policy liberalization.

Table 9.2. Regional Differences in Forest Resources

Region	Province		1977	1987
South	Jiangxi	Area	5.47	5.90
		Volume	298.60	242.10
	Fujian	Area	4.51	4.53
		Volume	430.35	396.65
North	Shandong	Area	1.60	1.60
		Volume	24.26	47.94
	Henan	Area	1.42	1.57
		Volume	68.22	91.52

Notes: Data are from the forest survey reports of the provincial forest services. Area is millions of hectares of closed forests. Volume is million cubic meters in both open and closed forests.

In the northern production region, policy reform clearly increased both short-term harvest activities and longer-term reinvestments in standing timber volumes. In the southern production region, household uncertainty with regard to long-term policy stability prevailed. Reform initially induced larger harvests but it failed to induce investments in timber management. Of course, subsequent southern policy recisions induced more harvesting on the remaining household lands, but also failed to induce investment. Standing forest inventories declined further and harvest levels began to fall as well—despite that fact that southern forests received subsidies and government investments for the remaining share of collective forests that were that still were not contracted to household management and also for the south's larger component of government forests.

The province of Anhui provides another comparison of these policy-induced effects on harvests and longer-term timber investments. The northern

1/3 of Anhui is an important agricultural area. Anhui's southern 2/3 lies in the southern timber production region. Timber harvests in both northern and southern parts of the province increased from 1977 to 1980, but southern harvest levels have declined ever since. Standing timber volumes in the north increased almost four-fold between 1977 and 1989, while they declined almost 10 percent in the south—despite the southern concentration of government timber subsidies and investments (Yong 1987).

The Production Assessment

One straightforward approach for assessing the impacts of China's policy reforms would be to regress short- and longer-term regional production responses (timber harvests and measures of the standing forest, respectively) on the two major policy instruments (HRS and price liberalization), and other factors as appropriate. The estimated policy parameters would indicate the effects of those instruments.

More formally:

$$\ln Y_{it} = \alpha_0 + \alpha_1 P_{it-1} + \alpha_2 HRS_{it-1} + \alpha_{3+g} D_g + \alpha_4 T + \epsilon_{it} \qquad (9.1)$$

where Y_{it} is harvest volume for region i at time t in the regression measuring short-run effects. Y_{it} is either standing timber volume or forest area in the regressions measuring longer-term effects. P_{it-1} is the timber price faced by farmers, deflated by the rural industrial producer price index; HRS_{it-1} is the ratio of contracted forest land to total collective forest land, T is a trend variable which captures any sequential effects contained in the dependent variables including technical change and input availability, and D is a vector of cross-sectional dummy variables. The subscripts i and t reflect the cross-sectional and intertemporal differences in our observations. Last year's procurement price and portion of land under contract are related to this year's production change in order to account for adaptive expectations and a lag in the production responses. ϵ_{it} is an independently distributed error term with zero mean and no intertemporal correlation. The α's are parameters to be estimated.

Data

Our regional data sets cover the period from 1978 to 1989 for four northern and five southern prefectures (Kaifeng, Zhoukou, Fuyang, Suxian; and Yichun, Ji-an, Fuzhou, Ganzhou, Nanping; respectively)—so that the northern data set has 48 observations and the southern has 60 observations. Local timber companies provided the data on harvest volumes and procurement and market prices. Surveys conducted by each province in roughly five-year intervals provided our measures of forest inventory and forest area.[7] We estimated intermediate annual values with the help of average growth rates for the regions. The provincial statistics services provided the rural industrial producer price index. Finally, the provincial forest agencies provided data on contracted and total collective forest land.

The 12-year time period of our data, our sample size, and the interpolations necessary to create our forest inventory and land area data raise technical econometric questions. Regarding the time period: Is it sufficient for the detection of long-run changes in forest production? While twelve years may be insufficient for identifying a new long-run equilibrium in forestry, we would argue that twelve years should be long enough to discern the sources of regional transitions. To justify this claim, we estimated average annual growth rates for our dependent variables (Table 9.3). In the north, timber harvests, stocking volume, and forest area grew at annual rates of 18.9, 7.5, and 11.2 percent, respectively. In the south in contrast, the same measures grew very slowly or even declined. (The north-south differences in inventory and area are statistically significant.) These differences seem to indicate substantially different regional production responses.

Table 9.3. Regional Forest Dynamics, 1978-89

	\multicolumn{6}{c}{Annual forest change in:}					
	Harvests		Inventory		Area	
	Mean value	Std. dev.	Mean value	Std. dev.	Mean value	Std. dev.
North	0.189	0.259	0.075	0.053	0.112	0.088
South	0.009	0.126	-0.010	0.006	0.008	0.009

Note: See the discussion for description of data.

We normalized each dependent variable in a prefecture by its 1980 level (to correct for differences in sample size) and we introduced the cross-sectional dummy variables. We can examine the potential effects of our interpolated forest inventory and forest area data by running three different equation forms: the full model and data set, the full model with data only from the three forest survey years, and a restricted model that drops the trend variable but uses all the data. We can be confident of our model specification and estimation if the observed differences in equation fits and the estimated coefficients are small.

Estimated Results

Tables 9.4 and 9.5 present our regressions for the north and the south, respectively. In general, the R^2 values indicate good equation fits in both regions. The changes in the R^2s are relatively small when we drop the trend variables—probably because the trend variables and the two policy variables (HRS and price liberalization) moved together during the period of our analysis. It is interesting to note that removing the trend variable causes the policy parameters to become larger (in all cases) for the north but smaller (in two cases of three) for the south. Therefore, the trend variables should constrain us from overstating policy effects in the north, or understating policy effects in the south.

The coefficient estimates themselves conform with expectations. In the north, harvests (columns 1 and 2) respond positively to both increasing prices and improved tenure. In the south, harvests respond positively (and significantly) to improved tenure but they display no significant relationship to increasing prices. This southern result shows that the prospect of loosing control of the resource dominates any prospect of increasing future prices. Therefore, farmers with uncertain tenure will harvest and sell what they can while they do control the resource, and regardless of future price prospects.

In the north, standing forest volume and forest area both (columns 3–4 and 5–6) respond positively, and significantly, to both policy variables. In the south the relationships between these measures of forest investment and the two policy variables are small, insignificant, (and generally negative) when we include the trend variable in the regression.[8] This is what we should expect if farmers harvest at their first opportunity and disinvest (harvest any subsequent growth without reinvestment) whenever permitted the renewed opportunity to harvest.

These regional differences support our contention that policy uncertainty works against long-term forest investment. That is, in the north, longer-term investments in forestry follow improved land tenure. Increasing prices act as further investment incentives once land tenure is more secure. In the south, the same two desirable policies, improved tenure and liberalized prices, fail to induce forest investment because farmers have no confidence that the policies will endure long-enough for them to gain the benefits of their investments.

Finally, the trend variable is always positive in the north, while it declines in the harvest and investment regressions for the south. We have explained how we expect the trend in southern investments to be downward in the absence of confidence about the security of household land tenure. With declining southern investments, the standing forest inventory must decline and this implies that future harvests must decline as well. The land area in forests

Table 9.4. Supply Response Regressions for the Northern Forest Production Region

	Harvests			Inventory			Area	
	(1)	(2)	(3)	(4)	(5)	(6)
Constant	-1.11***	-1.21***	-0.37***	-0.41***	-0.59***	-0.64***
	(3.82)	(3.77)	(4.99)	(4.32)	(4.83)	(4.53)
Price, P_{it-1}	0.29*	0.41***	0.13***	0.18***	0.17***	0.24***
	(1.93)	(2.36)	(2.98)	(3.39)	(2.51)	(3.06)
Tenure, HRS_{it-1}	0.003	0.009***	0.002***	0.005***	0.004***	0.007***
	(0.96)	(4.11)	(3.07)	(7.40)	(3.29)	(7.32)
Trend, T	0.10***		0.04***		0.05***	
	(3.19)		(5.14)		(3.81)	
Prefecture dummies						
D_1	0.03	0.01	-0.11***	-0.12***	-0.04	-0.06
	(0.03)	(0.01)	(3.44)	(3.01)	(0.77)	(0.94)
D_2	1.01***	0.95***	0.11***	0.09**	0.26***	0.23***
	(8.55)	(7.31)	(3.65)	(2.20)	(5.20)	(3.98)
D_3	0.52***	0.47***	0.13***	0.11***	0.28***	0.26***
	(4.37)	(3.60)	(4.26)	(2.84)	(5.62)	(4.46)
degrees of freedom	41	42	41	42	41	42
R^2	0.87	0.84	0.95	0.91	0.93	0.90

***, **, and * indicate statistical significance at the 0.01, 0.05, and 0.10 level, respectively. Asymptotic t tests shown in parentheses.

Table 9.5. Supply Response Regressions for the Southern Forest Production Region

	Harvests			Inventory			Area
	(1)	(2)	(3)	(4)	(5)	(6)
Constant	0.02 (0.37)	0.03 (0.42)	0.04*** (4.23)	-0.04*** (4.27)	-0.03*** (4.05)	-0.03*** (4.04)
Price, P_{t-1}	-0.09 (0.62)	-0.23* (1.84)	0.01 (0.32)	-0.03* (1.78)	-0.004 (0.18)	0.04*** (2.53)
Tenure, HRS_{t-1}	0.006** (2.27)	0.002 (1.38)	-0.001 (0.01)	-0.001*** (4.67)	-0.001 (1.08)	0.001*** (3.24)
Trend, T	-0.05* (1.81)		-0.01** (2.09)		0.01*** (2.67)	
Prefecture dummies						
D_1	-0.06 (1.05)	-0.05 (1.13)	-0.04*** (5.44)	-0.01*** (5.45)	0.01 (1.05)	0.01 (1.24)
D_2	0.07 (1.24)	0.06 (0.98)	0.01 (1.37)	0.01 (0.96)	-0.01 (1.22)	-0.01 (0.69)
D_3	0.05 (0.88)	0.03 (0.48)	-0.04*** (4.45)	-0.05*** (2.84)	0.04*** (4.65)	0.04*** (5.47)
D_4	0.35*** (5.91)	0.34*** (5.68)	-0.03*** (3.48)	-0.03*** (3.48)	0.02*** (2.96)	0.02*** (2.96)
degrees of freedom	52	53	52	53	52	53
R^2	0.55	0.52	0.81	0.79	0.77	0.74

***, **, and * indicate statistical significance at the 0.01, 0.05, and 0.10 level, respectively. Asymptotic t tests shown in parentheses.

should follow the same pattern as the standing forest inventory. That it does not is a reflection of government reforestation programs rather than private reinvestment. Unfortunately, increasing the forest area while decreasing the standing forest inventory means that the quality of the southern resource is also declining. Average stocking in the south is now about 43 m^3/ha., or less than half of the world average (Yang 1992).

We can use the estimated parameters from our equations to calculate contributions of the reform programs to economic growth. Comparing these policy-induced growth contributions with the contributions of similar agricultural policies should provide insight to the relative importance of the forest policy reforms as well as additional insight to the general reasonableness of our findings. There are four steps to the calculation: i) multiply the estimated coefficients by 100, ii) calculate the change in a variable as the weighted difference in magnitude of the variable among all cross-sections between the first and last sample periods, iii) find the product of results from the first two steps, and iv) compute the contributions of the residuals by dividing total growth by the degree of fit (R^2).

Table 9.6 shows this calculation and its results. In the north, price changes and tenure reform share approximately equally in 30 percent of the 184 percent growth in harvests between 1978 and 1989. Prices and tenure policy also explain approximately half of the increase in long-term forest investment in the north in this period (48 percent of the 95 percent growth in inventory and 47 percent of the 116 percent expansion in forest area). Comparable estimates for the agricultural sector show that rising prices and improved tenure account for 57 percent of the 42 percent growth in agricultural output from 1978 to 1984. Fifteen of the 57 percent was due to increasing agricultural prices while the remaining 42 percent was due to tenure reform (Lin 1992, McMillan et al. 1989, Wen 1994). Thus, the policy reforms induced a relatively greater expansion in forestry than in agriculture. This is an important, but unsurprising, observation because secure tenure should be a more important variable for longer-term investments and forestry is generally a longer term activity than agriculture.[9]

Table 9.6 also shows interesting results for the southern forest production region. Increases in southern harvests came almost exclusively from improvements in tenure. This observation is consistent with our previous reasoning that tenure reform dominates price increases in the presence of uncertainty. Managers must be confident of their control of the resource before price increases will induce them to invest. Table 9.6 also shows that the increases in forest area were due to neither price increases nor improved tenure. We previously suggested that these increases were due to government, rather than private, investment.

Table 9.6. Accounting for Forest Growth, 1978–89

Explanatory variable	Change in variable, 1978-79		Harvests		Inventory		Area
	(1)	(2)	(3)	(4)	(5)	(6)	(7)
NORTH							
Price	1.02	28.57	29.14 (15.84)	12.59	25.12 (26.51)	17.17	17.51 (15.11)
Tenure	89.35	0.28	25.29 (13.75)	0.23	20.64 (21.78)	0.41	36.45 (31.46)
Trend	11.00	9.67	106.37 (57.82)	3.98	43.78 (46.21)	4.89	53.79 (46.43)
Residual			23.18 (12.59)		5.21 (5.50)		10.51 (7.00)
Total growth			183.98 (100.00)		94.75 (100.00)		115.86 (100.00)

SOUTH

Price	1.66	−8.91	−14.79 (−142.70)	0.81	1.34 (13.90)	−0.38	−0.63 (−7.60)
Tenure	63.72	0.63	40.14 (387.50)	−0.09	−5.73 (−59.44)	−0.04	−2.55 (−30.79)
Trend	11.00	−4.54	−49.94 (−482.00)	−1.05	−11.55 (119.80)	1.07	11.77 (141.98)
Residual			14.23 (137.40)		6.30 (65.35)		0.30 (3.60)
Total growth			−10.36 (100.00)		−9.64 (100.00)		8.29 (100.00)

Notes: Column 1 is the change in the explanatory variable. Columns 2, 4, and 6 are the estimated coefficients (times 100). Columns 3, 5, and 7 are contributions to growth. Columns 3, 5, and 7 are the products from the multiplication of column 1 by columns 2, 4, and 6, respectively. The numbers in parenthesis are the percentage share contributions to total growth.

Discussion and Conclusions

Our analysis argues that forest production responses to policy instruments highlight a sharp contrast between the two regions. The contrasting production responses reflect the different policy environments and, therefore, the different producer expectations in the two regions. Reform policies have improved the incentive structure in the northern forest production region, but not in the south.

We can confirm this summary observation with evidence from surveys of household behavior, from a decomposition of the regional timber prices, and from a review of the physical forest resources in the two regions.

After our initial analysis was completed, we conducted household surveys of 100 farmers in each of the original northern and southern sampling areas. Most farmers in the north believed that the new tenurial arrangements and the open market for timber transactions work satisfactorily and should be maintained. The majority of southern farmers felt that their land tenure was insecure, and this insecurity contributed to their reluctance to plant and manage trees. In addition, approximately eighty percent of southern farmers felt that the controlled procurement prices facing them were low and that the burdens of taxes and fees imposed on them was heavy.

A decomposition of the 1990 timber prices (masson pine) for representative southern (Sanming, Fujian) prefectures lends substance to these farmer perceptions. It is unnecessary to decompose northern prices as farmers in this region receive market prices and hold onto over ninety percent of total revenues from their timber sales. In contrast, individual farmers and collectives in the south received only 31 percent of total revenues from their timber sales. The remaining 69 percent was consumed by taxes (23 percent), forest service revenues (26 percent), local government revenues (seven percent), and timber procurement company transaction costs (13 percent). These burdens on southern timber sellers decrease their incentives for forest production and support their perception that their timber prices are low.

We also examined the forest resource conditions in the two regions to see if physical factors constrained production. One could argue that the substantial increase in forested area in the north and the slight increment in the south might indicate that more land was available for tree-planting in the north. In fact, this was not the case. In the north, the percentage of land with forests and trees is more than 100 percent of the potential area available for forestry—because northern forests are well-stocked and various agroforestry regimes also put trees on some land that is primarily agricultural. In the south, however, forested land accounts for only 55 percent of the total land area potentially available for forestry. Furthermore, if the opportunities for forest management are dependent on the condition of existing timber stands, then it may not be practical to introduce additional management on stands that are well-stocked or close to maturity. In the south, however, about seventy percent of existing stands are understocked with young trees and the potential for inventory expansion is large (Ministry of Forestry 1988).

This additional information all supports our fundamental hypothesis that the effect of rural reforms critically depends on how the reforms are implemented. If they are implemented in a manner that limits the insecurities associated with land tenure and price expectations, then the reforms will improve the incentive structure and farmers will respond to the policy changes. Alternatively, if the reform policies are implemented in a manner that creates insecurity regarding land tenure, price distortions, and uncertain price expectations, then harvests will be pushed to unsustainable levels and forest growth and development will be limited.

In sum, China's contrasting regional experiences illustrate the critical importance of a stable policy environment as a vehicle for economic growth. Without a stable policy environment, positive incentives contribute little to producers' longer-term expectations and farmers will reject opportunities to commit resources to longer-term activities. Producers will take short-term advantage of any new incentive and disregard uncertain longer-term opportunities. In forestry, producers will harvest in the short run without replacement until eventual longer-term timber production must also decline. Therefore, stable policy environments and sustainable growth are closely related, and forestry development around the world has much to gain from these observations on China's rural reforms.

NOTES

1. University of Georgia, University of Georgia, and Centre for International Forestry Research, respectively. A prior version of this paper appeared as "Impacts of rural reforms: the case of the Chinese forestry sector," *Environment and Development Economics* 2(3):289–304.

2. Actually, household responsibility replaced township responsibility. By the late 1970s, townships had succeeded production teams as the focal production unit.

3. The SCFR consists of Hubei, Hunan, Fujian, Zhejiang, Guangdong, Guangxi, Hainan, Guizhon, Jiangxi, and parts of Sichuan, Yunan, Anhui, and Jiangsu. The eastern CNNFFR consists of Hebei, Liaoning, Henan, Shandong, Beijing, Tianjin and parts of Anhui, Jiangsu, Shanxi, and Shaanxi.

4. In a few places, notably northwest Fujian province, an organizational form known as a shareholding corporation emerged. Shareholding corporations converted all collective commercial forests into monetary shares and distributed the shares to individual households. To date, shareholding corporations perform much like the old regime—with only a name change (Sun 1992).

5. Our measure includes agricultural lands and all collective forest lands. State forests are excluded. State forests are more important in the southern forest productive region where they account for about ten percent of the total forest area.

6. The merits of secure land tenure are known to forestry (Mendelsohn 1994) and they have been demonstrated empirically for agriculture (Feder et al. 1988, Besley 1995). It is the general uncertainty, not the change in tenure, that makes the experience of China's southern forest production region an interesting case.

7. China conducted three nationwide surveys within the general period of our analysis. In general, the first was completed in 1976, the second in 1981 and the third in 1987.

8. In addition, the coefficient of each policy instrument in one region is significantly different from the comparable coefficient in the other region.

9. Furthermore, our estimate of the effect of secure tenure for this northern region is conservative because HRS reform was largely complete by 1984, but our data cover the longer period until 1989. This makes our trend coefficient more prominent relative to our coefficient on HRS.

REFERENCES

Besley, T. 1995. "Property rights and investment incentives: Theory and evidence from Ghana." *Journal of Political Economy.* 103:903–937.
China Agricultural Yearbook, 1990–92. Beijing: China's Agricultural Press.
Deacon, R. T. 1994. "Deforestation and the rule of law." *Land Economics.* 70:414–430.
Feder, G., T. Onchan, Y. Chalamwong, and C. Hongladarom. 1988. *Land Policies and Farm Productivity in Thailand.* Baltimore: Johns Hopkins University Press.
Li, J. C., F. W. Kong, and N. H. He. 1988. Price and policy: "The key to revamping China's forestry resources." In R. Repetto and M. Gillis, ed., *Public Policies and the Misuse of Forest Resources.* Cambridge: Cambridge University Press.
Lin, Y. J., 1992. "Rural reforms and agricultural growth in China." *American Economic Review* 81:34–51.
McMillan, J., J. Whalley, and L. J. Zhu. 1989. "The impact of China's economic reforms on agricultural productivity growth." *Journal of Political Economy* 97:781–807.
Mendelsohn, R. 1994. "Property rights and tropical deforestation." *Oxford Economic Papers* 46:750–756.
Ministry of Forestry. 1988. *Resource Bulletin.* Beijing.
Perkins, D., and S. Yusef. 1984. *Rural Development in China.* Baltimore: Johns Hopkins University Press.
Sun, C. J. 1992. "Community forestry in southern China." *Journal of Forestry* 90:35-39.
Wang, Y. C., D. S. Liu, and J. T. Xu. 1991. *A study of forestry development in the South.* In W.T. Yong, ed. Studies on China's Forestry Development. Beijing: China's Forestry Press.

Wen, J. G. 1993. "Total factor productivity growth in China's farming sector: 1952–1989." *Economic Development and Cultural Change*. 41:1–41.

World Bank. 1985. *China: Agriculture to the Year 2000*. Washington.

Yin, R. S. 1994. "China's rural forestry since 1949." *Journal of World Forest Resource Management*. 7:73–100.

Yin, R. S. and J. T. Xu. 1987. "A survey of timber revenue distribution before and after opening the market." *Forestry Problems* 1:109–128.

Yong, W. T. 1987. "How to enhance the forest productivity." *Forestry Problems*. 1:19–40.

Zhong, M. G., C, Xian, and Y. M. Li. 1991. "A study of forestry development in the major agricultural areas." In W.T. Yong, ed., *Studies on China's Forestry Development*. Beijing: China's Forestry Press.

Part 6
Additional Perspective

Our first two analytical chapters examined a) migration to forested uplands and b) the most general expectations and preferences of farm households regarding trees. These chapters were intended to indicate some of the breadth of social forestry as a topic of scientific inquiry and policy analysis. Subsequent chapters have been narrower and more technical. Perhaps it is now time to step back and reconsider the breadth of social forestry, and to consider what we may have overlooked.

Two items come to mind: common property, and especially its importance to the poorest of the poor, and the potentially beneficial impacts of trees on other human activities. These are the topics of the next two chapters. Both are topics of widespread discussion. Neither has been the focus of much empirical economic analysis.

N.S. Jodha is the widely recognized expert on common property resources and rural poverty. In chapter 10, he provides a wealth of evidence from Rajasthan in India supporting the contention that the poor are unusually dependant on common resources for wood and other resource services, and that this dependence may be increasing despite regulations on resource use and land redistribution. Jodha's evidence begs our contemplation of what happens to the poor as resources become scarcer and as more secure claims arise for forest resources that previously were open to access by all local user groups.

Chapter 11 uses an example from agroforestry in China to address new questions about the contribution of forest-based activities to sustaining the local environment and its potential for long-term productivity. Agroforestry becomes important as forest resources become scarce enough to justify farmers using their already scarce land and labor inputs to grow trees. Farmers may grow trees for their products or for their positive interactive effects on other agricultural crops. The idea of positive interactive effects is at the core of agroforestry. The north central China case discussed in chapter 11 measures eight percent gains in grain production due to the erosion control and organic matter contributions of intercropped trees. This evidence encourages a belief in the economic role of trees and related environmental improvements to sustain agriculture production.

CHAPTER 10

Common Property Resources and the Dynamics of Rural Poverty: Field Evidence from the Dry Regions of India

N. S. Jodha[1]

Despite advances in agricultural technologies, access, and the availability of external inputs and supplies, and despite income transfers through subsidies and relief, locally available biomass continues to be an important element in the fulfillment of the basic needs of rural people, and especially the rural poor. In addition to meeting the direct needs of fodder, fuel, food, timber, fencing, thatching, etc., biomass plays a crucial role in the local resource regenerative processes of farming systems. These benefits notwithstanding, the supplies and sources of biomass are rapidly being depleted in the dry tropical regions of India. Farm-produced supplies suffer due to the anti-biomass bias of new crop technologies favouring high grain-stalk ratios and short duration cropping systems, and due to the emphasis on cash crops (Jodha 1991). Furthermore, natural supplies of biomass are also declining due to decreases in land areas in natural vegetation, as well as to decreases in their productivity. The heaviest burden of these developments falls on the rural poor who do not have enough other options to meet their basic requirements for these products. The changing situation for the common property resource (CPRs) on which the rural poor depend illustrates each of these points.

Premises

The key issues for CPRs and the dynamics of rural poverty are as follows:

1. In the absence of alternative income and employment generating options, the fate of the rural poor and the CPRs are interlinked.
2. CPR-rural poor interactions are an important dimension in the dynamics of rural poverty for the dry regions of India. The degradation of CPRs, and the ever-increasing dependence of the poor on CPRs, represent an invisible process of pauperisation. Non-recognition, and non-enumeration by the statistical system,

of the CPRs' contributions and the poor's dependence on them, on the one hand, and the failure of fiscal systems to include the changing status of CPRs and their contributions, on the other, are key factors responsible for this invisible pauperisation.
3. The dynamics of rural poverty, using CPRs as key vector, can be examined at three levels:
   (a) The changing status of CPRs in terms of their area as well as their physical productivity, and the changing management of the former.
   (b) The changing economic situation of groups depending heavily on CPRs for the bulk of their sustenance.
   (c) The place of CPRs in the anti-poverty programmes of the government.

This chapter is based on a study of common property resources in dry regions of India that covers over 80 villages in over 20 districts of 6 states. Methodological details and more complete evidence on the issues discussed here are presented elsewhere (Jodha 1986, 1990a,c, 1992). This chapter first describes CPRs and their contributions, and then presents village level evidence on the dependence of poor households on CPRs. A third section of the chapter comments on the decline of CPRs and the factors explaining it. The final section examines public interventions involving rural poor and CPRs.

Dependence of the Rural Poor on Common Property Resources

CPRs are non-exclusive resources. Members of the community share in the use rights and obligations for these resources (Bromley and Cernea 1989, Magrath 1986, Ostrom 1988). In simple terms, CPRs could be described as those of a community's natural resources to which every member of the community has access and the opportunity for its use. Members of the community have definite obligations but no exclusive property right to the resource. In the dry tropical regions of India, CPRs include community forests, pastures, common dumping and threshing grounds, rivers and rivulets, their banks and beds, watershed drainage, village wastelands, etc.

Legally, some CPRs may belong to other agencies (*e.g.*, the village wastelands may belong to the revenue department of the state); but they are under the *de facto* use and management of village communities. Despite differences in some of their specific uses and in their legal nomenclature in village revenue records, most CPRs are less separable interms of: (a) utility—as sources of varied biomass; (b) their current vegetative make-up consisting of sparse trees, shrubs, grasses and empty patches, and the seasonality of their product flows; (c) the usage regulations or, rather, disregard for them, and the consequent pace and pattern of CPR degradation; and (d) the indiscriminate coverage of most of them in public programmes like social forestry. Hence, one can use the term "forest-CPRs" without implying restriction to community forests. The term may encompass all CPRs contributing to the natural biomass supplied to village communities.

Figure 10.1. Districts and Numbers of Villages Covered by the Study on Common Property Resources in Dry Regions of India

Potential Gains

Despite rapid decline in their area and productivity, CPRs constitute an important component of community assets in the dry areas of India and other developing countries (Sandford 1983, Bromley and Cernea 1989, Margrath 1986, Ostrom 1988, Feeny, 1990). In dry areas, CPRs provide one of the community's responses to the scarcities and stresses created by agro-climatic conditions. These conditions call for diversified farming systems involving combinations of annuals and perennials, complementary activities based on CPRs and private property resources, and group action for risk sharing and resource management (Jodha 1992). CPRs are sources of a bulk of physical products, they offer employment and income generation opportunities, and they provide larger social and ecological gains.

Table 10.1 provides an indicative picture. It reveals that CPRs can contribute several elements to anti-poverty programmes if they are recognized as a part of development strategies. On the other hand, it is not difficult to visualize a process of pauperisation associated with decline in the status and productivity of CPRs, and the consequent loss of their potential contributions. Such declines are especially important for the poor who depend on CPRs more than the rich. Ironically, most of the potential contributions of CPRs indicated in table 10.1 are also goals of several public interventions—but these goals are rarely understood in the CPR perspective.

Dependence of the Poor

Table 10.2 highlights some observations of the gains from CPRs. CPRs have been degraded and their productivity is much lower today than in the past. Consequently, the rural rich (large farmers indicated by the "others" category in table 10.2), depend very little on CPRs. It is not worthwhile for them to collect and use meager quantities of products from these resources. The rural poor (small farmers and landless labourers) with limited alternative opportunities, increasingly depend on low pay-off options offered by CPRs. In the villages of this study, 84–100 percent of the rural poor depended on CPRs for fuel, fodder, and food. The corresponding proportion of rich farmers did not exceed 20 percent (except in very dry villages of Rajasthan). Intermediate categories of farm households (not shown in the table 10.2) depended on CPRs more than the rich.

Table 10.2 reveals that: (a) The rural poor satisfy 66–84 percent of their fuel requirements and 79–84 percent of their (fodder) grazing requirements from CPRs. (b) CPR collection provides 136–196 person-days of employment per household per year. CPR-related income accounts for 17–23 percent of total household income from other sources. (c) Inclusion of

CPR income in total household income reduces the extent of rural income inequity, as measured by the reduced value of the Gini-coefficient for income distribution. Data presented elsewhere (Jodha 1992) also suggest that CPRs not only provide input (in cash or kind) for farm operations during normal years, but contribute 25 to 38 percent of sustenance income (excluding credit) during drought years.

Table 10.1. Contributions of Common Property Resources to Village Economies in Dry Regions of India[a]

Contributions	CPRs[b]					
	A	B	C	D	E	F
Physical products						
food/fibre items	x		x	x		
fodder/fuel/timber/etc.	x	x	x		x	x
water				x	x	
manure/silt/space	x	x	x			x
Income/employment gains						
off-season activities	x				x	x
drought period sustenance	x	x				x
additional crop activities			x	x		x
additional animals	x	x				
petty trading/handicrafts	x					x
Larger social, ecological gains						
resource conservation	x	x		x		
drainage/recharge of ground water			x		x	
sustainability of farming systems	x	x	x		x	x
renewable resource supply	x	x	x			
better micro-climate/environment	x	x		x	x	

[a] Table from Jodha (1985a)
[b] CPRs: A - Community forest; B - Pasture/waste land; C - Pond/tank; D - River/rivulet; E - Watershed drainage/river banks; F - River/tank beds

Thus, the heavy dependence of the rural poor on CPRs links these resources to the dynamics of poverty and to poor-centred development interventions. Therefore, any change in the status and productivity of CPRs directly influences the economy of the rural poor.

Depletion of CPRs

The indifference, insensitivity, and outright negative approaches of the state and its development interventions for CPRs suggest that neither sustenance of

Table 10.2. Extent of Dependence on CPRs in the Dry Regions of India[a]

States (nos. of districts and villages)	Household categories[b] (%)	CPR Contributions to Household Consumption, Employment, Income, etc.							
		Fuel supply[c] (%)	Animal grazing[d]	Per household			Income Gini coefficient[h]	
				Employment[e] (no. of days)	Annual income[f] (Rs.)	CPR-income[g] (%)	Income from all sources	All sources excluding CPRs (%)
Andhra Pradesh (1,2)	Poor	84	-	139	534	17	0.41	0.50
	Others	13	-	35	62	1	0.41	0.50
Gujarat (2,4)	Poor	66	82	196	774	18	0.33	0.45
	Others	8	14	80	185	1	0.33	0.45
Karnataka (1,2)	Poor	-	83	185	649	20		
	Others	-	29	34	170	3		
Madhya Pradesh (2,4)	Poor	74	79	183	733	22	0.34	0.44
	Others	32	34	52	386	2	0.34	0.44

State	Group							
Maharastra (3,6)	Poor	75	69	128	557	14	0.40	0.48
	Others	12	27	43	177	1	0.40	0.48
Rajasthan (2,4)	Poor	71	84	165	770	23		
	Others	23	38	61	413	2		
Tamil Nadu (1,2)	Poor	-	-	137	738	22		
	Others	-	-	31	164	2		

[a] Based on village/household data reported in Jodha (1986).
[b] Number of sample household from each village varied from 20 to 36 in different districts. 'Poor' include agricultural labourers and small farm (<2 ha. dryland equivalent) households. 'Others' include large farm households only.
[c] Fuel gathered from CPRs as proportion of total fuel used during three seasons covering the whole year.
[d] Animal unit grazing days on CPRs as proportion of total animal unit grazing days.
[e] Total employment through CPR product collection.
[f] Income mainly through CPR product collection. The estimation procedure underestimates the actual CPR income. (See Jodha 1986.)
[g] CPR income as % of income from all sources.
[h] Larger Gini coefficients indicate greater degrees of income inequality. Calculations based on income data for 1983–84 from a panel households in ICRISAT's village level studies (Walker and Ryan 1990). The panel of 40 households from each village included 10 households from each of the categories, large, medium, and small farm households, and labour households.

the poor and contributions to private resource-based farming nor the ecological gains from CPRs seem to be a part of state rural development strategies (Jodha 1992). The result is reflected in the rapid decline in the area of CPRs, in their physical degradation, and in the minimal public effort expended for their development and regulated use. Detailed evidence from the study villages provides supports these arguments.

Table 10.3 shows that since the early 1950s when land reforms were introduced in most parts of the country, the area of CPRs has declined by 31–55 percent in the study villages. Other studies corroborate this observation (Iyengar 1988, Blaikie et al. 1985, Brara 1987, Oza 1989, Chopra et al. 1990, Chen 1988, Arnold and Stewart 1990). The pressure on remaining CPRs has increased rapidly as a combined result of the reduced area in CPRs and population growth. For instance, the average number of persons per 10 hectares of CPR ranged from 13 to 101 in 1951. By 1982, the same measure increased to the range of 47–238, depending on the sample village.

The immediate consequence of increased pressure on CPRs is their over-exploitation and degradation (table 10.4). The physical degradation of CPRs is strongly felt and observed, but its quantification is difficult due to the lack of benchmark data. Nevertheless, case histories and close monitoring provides the basic details. Declines in the number of products available and their yields are the main indicators of physical depletion. For instance, the number of different CPR products collected by villagers ranged from 27 to 46 before 1952. At present, the range is between eight and 22 products. The decline in the number of CPR products also indicates the reduced biodiversity maintained in CPRs. In sum, there has been a decrease in both the range and the diversity of production and consumption options for the rural poor.

Besides overcrowding, an important reason for the degradation of CPRs is a slackening of traditional management. State interventions have been ineffective in substituting formal systems for the previous informal social sanctions and customary arrangements for protecting, upgrading and regulating the use of CPRs. As a result, many CPRs have become open access resources, with everyone using them without any reciprocal obligation to maintain them. Table 10.5 shows that nearly 90 percent of villages presently fail to enforce historic formal and informal regulations. The corresponding percentages of villages failing to collect formal and informal user fees for CPR maintenance, and of villages that do not enforce other obligations for CPR upkeep, are 100 and 84 percent, respectively. The consequences are overexploitation and rapid degradation.

Reduction in land area, poor maintenance, and the decline in carrying capacity are clear indicators of reduced supplies of products for those who depend on CPRs. Seen in relation to earlier evidence of the heavy dependence of the rural poor on CPRs, the decline of the CPRs represents a further step

Table 10.3. Extent and Decline of CPR Area[a]

States (no. of districts)	Study villages (no.)	Area of CPRs 1982-84[b] (ha)	CPRs as proportion of total village area		Decline in the area of CPRs since 1950-52 (%)	Persons per 10 ha of CPR	
			1982-84 (%)	1950-52 (%)		1950-52 (no.)	1982-84 (no.)
Andhra Pradesh (3)	10	827	11	18	42	48	134
Gujarat (3)	15	589	11	19	44	82	238
Karnataka (4)	12	1165	12	20	40	46	117
Madhya Pradesh (3)	14	1435	24	41	41	14	47
Maharastra (3)	13	918	15	22	31	40	88
Rajasthan (3)	11	1849	16	36	55	13	50
Tamil Nadu (2)	7	412	10	21	50	101	286

[a]Table adapted from Jodha (1986), where more disaggregated details are reported.
[b]CPRs include community pasture, village forest, waste land, watershed drainage, river and rivulet banks and beds, and other common lands. This statistic indicates average area per village.

Table 10.4. Indicators of Physical Degradation of CPRs[a]

Indicators of changed status and context for comparison	Andhra Pradesh (3)	Gujarat (4)	Karnataka (2)	Madhya Pradesh (3)	Maharashtra (3)	Rajasthan (4)	Tamil Nadu (2)
Nos. of CPR products collected[b]							
in the past	32	35	40	46	30	27	29
at present	9	11	19	22	10	13	8
No. of trees and shrubs per ha. in:							
protected CPRs[c]	476	684	662	882	454	517	398
unprotected CPRs	195	103	202	215	77	96	83
No. of watering points (ponds) in grazing CPRs							
in the past	17	29	20	16	9	48	14
at present	4	13	4	3	4	11	3
No. of CPR plots where rich vegetation, indicated by its nomenclature, is no longer available	-	12	3	6	4	15	-
CPR area used for cattle grazing in the past, currently grazed mainly by sheep/goat (ha)[d]	48	112	95	-	52	175	64

[a] The choice of CPRs was guided by the availability of past information about them. Past information collected from oral and recorded description of CPRs in different villages in the period preceding the 1950s. Current status (1982-84) based on observation and physical verification.
[b] Includes different types of fruits, flowers, leaves, roots, timber, fuel, fodder, etc. in the villages.
[c] Protected CPRs are the areas (called 'oran', etc.), where, for religious reasons, live trees and shrubs are not cut. These CPR plots (numbering between 2 to 4 in different areas) were compared with bordering CPR plots of CPRs which were not protected by religious or other sanctions.
[d] Area covered by specific plots that were traditionally used for grazing high productivity animals (e.g., cattle in milk, working bullocks, or horses of feudal landlords). Because of CPR depletion, these animals are no longer graze here.

towards pauperisation of the poor. They are left with no alternatives. The poor depend on increasingly inferior and insecure options offered by degrading CPRs. In the process, they over-extract from the CPRs and end up with still more inferior options. This is a classic case of the vicious circle of poverty and resource degradation reinforcing each other. There may be few parallels to this invisible process of simultaneous pauperisation of the community and degradation of its resource base.

Table 10.5. Indicators of Changes in Management of CPRs[a]

States (no. of villages)	Formal/informal regulations on CPR use[b]		Formal/informal taxes/levies on CPR use[d]		Users' formal/informal obligation for CPR upkeep[e]	
	In the past[c]	At present[c]	In the past	At present	In the past	At present
Andhra Pradesh (10)	10	-[f]	7	-	8	-
Gujarat (15)	15	2	8	-	11	2
Karnataka (12)	12	2	9	-	12	3
Madhya Pradesh (14)	14	2	10	-	14	3
Maharastra (13)	11	1	6	-	10	1
Rajasthan (11)	11	1	11	-	11	2
Tamil Nadu (7)	7	-	4	-	7	1
Total (%)	100	11	100	00	100	16

[a] Table based on field work during 1982-84 (Jodha 1986).
[b] Measures such as regulated/rotational grazing, seasonal restrictions on use of CPRs, provision of CPR watchmen, etc.
[c] 'Past' stands for period prior to the 1950s. Present stands for the early 1980s.
[d] Measures such as grazing taxes, levies, and penalties for violation of regulations on use of CPRs. See Jodha (1985b) for a descriptive account.
[e] Measures such as contribution towards desilting of watering points, fencing, trenching, protection, etc.
[f] (-) indicates nil.

These results are unnecessary, however, provided the poor are offered alternative options that would reduce their dependency on the common resources, and that would regulate the usage of and enhance the regeneration and raise the productivity of the CPRs. The next section of this chapter considers the search for such relevant alternatives for the dry regions of India.

CPRs and the Rural Poor: The State Approach

It would be wrong to conclude that public policies in the dry regions of India are completely insensitive to the problems of rural poor. In fact, ever since the initiation of economic planning in 1950s, the state has undertaken measures to help the poor. The focus of this chapter does not permit enumerating these measures. Rather, we will restrict comment to those anti-poverty measures of the state where CPRs are either directly or indirectly involved. The major thrusts of public policies involving the rural poor and CPRs could be put under the following categories: (a) asset redistribution, (b) productivity promotion, (c) formal management systems, and (d) biomass production projects. We might explore the role of CPRs in each.

Asset (Land) Redistribution

Redistribution to the landless and to small landowners was the key element of the land reforms introduced in the early 1950s in India. Having failed to acquire surplus land from large farmers and absentee landlords through effective land ceiling laws (Ladejinsky 1922), the state governments found it easier to redistribute CPR lands. This was done both by regularizing illegal encroachments as well as by formal redistribution of CPR lands. Table 10.3 demonstrated the consequences (the rapid decline of CPR area) of this undeclared state policy against CPRs. While most CPR lands are fragile, submarginal, and best-suited to natural vegetation, their privatization immediately put them under the plough. Therefore, one consequence of CPR privatization was low and unstable crop yields. Grain yields from ex-CPR lands have been one-fourth to one-half the yields on traditionally cropped lands. Such low grain yields do not compensate for the loss of biomass produced on these lands in the past (Jodha 1992).

A second and more serious aspect of the privatization of CPRs is the huge gap between the intention (land to the landless and the poor) and the reality of land distribution. Table 10.6 shows the details of the share of the rural poor among the total beneficiaries of CPR privatization. Notwithstanding slogans to the contrary, a large proportion of the ex-CPR land went to the non-poor, although the number of non-poor beneficiaries was fewer than the number of poor beneficiaries. On an average, non-poor families received two to three hectares of additional land. The average area received by poor households was barely one hectare. The last column of table 10.6 shows that those who already had more land also received more of the privatized CPR

Table 10.6. Distribution of Privatized CPR Lands to Different Household Groups[a]

(1) States (no. of villages)	(2) Total area redistributed (ha)	(3) No. of households receiving land	(4) Share of poor[b] in (2) (%)	(5) Proportion of poor in (3) (%)	Land received per household by:		Average area per household after receiving new land	
					Poor (ha)	Others[c] (ha)	Poor (ha)	Others (ha)
Andhra Pradesh (6)	493	401	50	74	1.0	2.1	1.6	5.0
Gujarat (8)	287	166	20	45	1.0	2.6	1.8	9.4
Karnataka (9)	362	203	43	65	1.3	3.0	2.2	8.0
Madhya Pradesh (10)	358	204	42	62	1.2	3.2	2.5	9.5
Maharashtra (8)	316	227	38	53	1.1	1.9	2.0	6.2
Rajasthan (7)	635	426	22	36	1.2	3.2	1.9	7.2
Tamil Nadu (7)	447	272	49	66	1.0	1.5	1.9	6.7

[a] Table adapted from Jodha (1986). Villages are from three districts in each state except for Andhra Pradesh (2) and Tamil Nadu (2).
[b] 'Poor' includes agricultural labourers and small farm (<2 ha of dry land equivalent) households.
[c] 'Others' in this table, include both medium and large farm households.

land. Furthermore, since the newly received land was too poor, unproductive and difficult to develop without the sort of complementary resources that were unavailable to the rural poor, 23 to 45 percent of the poor households dispossessed their new CPR lands (Jodha 1986).

In sum, it is doubtful whether the collective loss of the rural poor as former major users of CPRs was balanced by their individual gains as new land owners of ex-CPR lands. Thus, the first anti-poverty intervention involving CPRs did not help the poor.

Productivity Promotion

The government was alarmed by the physical degradation and falling productivity of village forests, community pastures, etc. It attempted several measures to raise their productivity. A number of research institutions and pilot development projects set to this task. However, while doing so, community resources were treated merely as pieces of physical resources located in the villages. Public interventions did not involve the local people or the CPR-perspective (represented by the relationship of a community to its natural resources). The village Panchayats (elected village councils) most of whom represented the state authority at the village level were involved, although ineffectively.

Public initiatives focused on techniques, rather than users' needs, and on means, as reflected by emphasis on choice of species, planning methods, reseeding techniques, scientific monitoring of plant coverage, etc. Most involved limited and selected species, at times exotic ones, which failed to meet the mixed biomass needs of the people. Furthermore, many programmes placed restrictions on local people's access to their own common resources. For example, pilot projects focussed on demonstrating potential technologies under ideal (user-less) situations. User perspectives and input from the village community were not emphasized. People's participation was not a strong point of these programmes, despite loud slogans to the contrary.

Without local participation, successful implementation of these programmed was difficult. They were (and continue to be) sustained by state grants in select pilot project areas, but most efforts to upgrade CPR productivity have proved irrelevant or ineffective (Gupta 1987, Shankarnarayan and Kalla 1985, Jodha 1988).

Formal Management Systems

CPRs represent one of the traditional forms of rural cooperation. They involve group activities guided by social sanction and convention, and enforced by the authority of village elders or, in some cases, feudal landlords. This helps in protecting, developing and regulating use of CPRs (Jodha 1985b). The feudal system was abolished with the introduction of land reforms in the early 1950s. As a part of next democratic programme, the elected village councils (Village Panchayats that replaced the traditional informal arrangements) were given responsibility to administer and implement development and welfare activities at the village level. Management of CPRs also became their responsibility. Despite all legal provisions, however, the Panchayats generally failed to undertake measures for CPR management. These would have annoyed their electorates. Panchayats often confined their roles to securing government grants in the name of CPRs and using them elsewhere (Jodha 1990c). Consequently, even the traditional management practices reported in table 10.5 were discontinued in most villages. Common property resources became open access resources, inviting the well-known "Tragedy of Commons." In sum, the usurpation of community mandates and initiatives by the state through a variety of legal, administrative and fiscal measures, further marginalised the role of communities in managing their common resources.[2]

Biomass Production Projects

The latest phase in the public approach to CPRs is represented by programmes for natural biomass production on community lands, including most of the CPRs.

The state was alarmed by the emerging biomass crisis, pushed by environmental lobbies, induced by donor recommendations, and encouraged by the achievements of small scale, scattered, NGO-supported initiatives, and workable scientific recommendations. In response, it has worked through several welfare and production programmed and resource development projects to initiate a number of activities to enhance biomass availability for village communities. Social and community forestry projects and integrated watershed management projects are examples. Despite occasional successes, these programmes have yet to make significant impacts on the rehabilitation of CPRs, or on the rural poor (Chambers et al. 1989). Except for a few operated by local NGOs and enlightened village Panchayats, most programs continue to share the features of past interventions. For instance, most of these programs still operate in project mode, are sustained by state subsidies and managed by state administrative or technical agencies, and most remain technique dominated and without sufficient people's participation. The very scale and magnitude of the problem is a big constraint. Inadequate understanding of the CPR-dimension of these resources is another problem.

CPR-Poverty Link: the Invisible Dynamics

The result of these various state programs is that the rural poor in most areas continue to depend on rapidly shrinking and depleting CPRs. An invisible process of pauperisation (Jodha 1990c) develops as the costs of production (largely the time of the rural poor) from CPRs increase while the outputs of CPRs (tables 10.1 and 10.2) decrease. The overall range and quality of options (the variety and quality of products) are declining (Jodha 1985b, 1992). The decline of the CPRs reflects various dimensions of the dynamics of rural poverty.

First, the transfer of submarginal CPR lands to private crop cultivation represents a step towards long-term unsustainability in dry areas. Private agricultural production ensures only a meager grain output, but it imposes a huge cost in terms of ecologically better-suited products (i.e., biomass) which would help sustain diversified farming (Jodha 1991). The poor suffer the most severe consequences.

Second, the reduced range and quality of employment and income options for succeeding generations of those dependant on CPRs will widen inter-generational inequity. This is a key element of the unsustainability phenomenon. One manifestation of it can be seen in premature harvesting and lopping of trees to make up for the reduced availability of plant material (Jodha 1993).

Third, the willing choice of CPR users for inferior options is another manifestation of unsustainability. Despite the reduced quantity and quality of products, the rural poor increasingly depend on CPRs because the opportunity cost of poor people's labour is still lower yet. This process is part of the progressive pauperisation of the rural masses (Jodha 1990c).

The whole process remains formally invisible, as it not reflected in the national accounts and it has no fiscal dimension (subsidy, relief, etc. for the poor). The community silently eats away its permanent natural assets and, in the final analysis, the loss of CPRs may prove more costly than alternative measures to help the poor.

Finally, the ultimate consequence of CPR degradation may be the permanent disruption or elimination of vital bio-physical processes, of nature's regenerative activities, (energy and material flows etc.), inside and even outside the CPR area within the overall watershed (Jodha 1991). These disruptions may further reduce the efficacy of farmers' traditional adaptation strategies against environmental stresses in dry regions (Jodha 1990b).

Thus, with reference to biomass production and other vital ecological parameters, and also with reference to the sustenance of the poor, the depletion of CPRs generates a process of scarcities and degradation that reinforce each other. This may be called the 'vicious circle of poverty and degradation'.

The situation is depressing, but it may become still more desperate if positive CPR policies are not adopted. The key elements of such policies are elaborated upon elsewhere (Jodha 1992). We can be indicate them here briefly:

1. Deliberate public action is required to restrict further curtailment of CPRs. Public welfare and development programmes need to be sensitized to CPR issues. General development policies intended to improve productivity and environmental stability can be made more successful if they are reoriented with a CPR perspective.
2. The key issues for protecting and rehabilitating CPRs are to reduce their mining by the users (due to the degradation-poverty cycle) and to introduce technological investments to raise CPR productivity.
3. The regulation of CPR use is equally important. This calls for the involvement of user groups and the mobilization of a community strategy that complements state interventions with the essential participation of local people. Recent experiences of successful participatory natural resource management initiatives can offer useful lessons for replication (Mishra and Sarin 1987, Chopra et. al. 1990, Shah 1987, Agrawal and Narain 1990, Campbell and Denholm 1992).

Increasing the visibility of CPR's contributions can help to mobilize policy and programme support. The dependence of the poor on CPRs, and the vicious circle of poverty and CPR degradation is an important dimension of the dynamics of rural poverty.

NOTES

1. Head, Mountain Farming Systems Division (MFS), International Centre for Integrated Mountain Development (ICIMOD), Kathmandu, Nepal at the time this paper was written.
2. More recently, a few initiatives supported by NGOs are attempting to restore local community control over local resources (Oza 1989, Shah 1987).

REFERENCES

Agarwal, A., and S. Narain. 1990. *Towards Green Villages*. New Delhi: Centre for Science and Environment.

Arnold, J.E.M., and W.C. Stewart. 1990. "Common property resource management in India." London: Oxford Forestry Institute paper 24.
Blaikie, P.M., J.C. Harriss, and A. N. Pain. 1985. "The management and use of common property resources in Tamilnadu." London: Overseas Development Administration.
Brara, R. 1987. "Shifting sands: a study of rights in common pastures." Jaipur: Institute of Development Studies.
Bromley, D.W., and M.M. Cernea. 1989. "The management of common property natural resources: some conceptual and operational fallacies." Washington: World Bank discussion paper 57.
Campbell, G., and J. Denholm. 1992. "Inspirations in community forestry." Kathmandu: ICIMOD, Report of the seminar on Himalayan community forestry.
Chambers, R., N.C. Saxena, and T. Shah. 1989. *To the Hands of the Poor: Water and Trees*. New Delhi: Oxford and IBH Publishing Co. Pvt. Ltd.
Chen, M. 1988. "Size, status and use of common property resources: a case study of Dhevdholera village in Ahmedabad district, Gujarat." Paper presented at women and agriculture seminar. Trivendrum, Kerala: Centre for Development Studies.
Chopra, K., G.K. Kadekodi, and M.N. Murty. 1990. *Participatory Development: People and Common Property Resources*. New Delhi: Sage Publications.
Gupta, A.K. 1987. "Why poor people do not cooperate? A study of traditional forms of cooperation with implications for modern organization." In G.C. Wanger, ed., *Politics and Practices of Social Research*. London: George Allen and Unwin.
Harriss, B., S. Guhan, and R. H. Cassen (eds). 1992. *Poverty in India: Research and Policy*. Bombay. Oxford University Press.
Iyengar, S. 1988. "Common property land resources in Gujarat: some findings about their size, status and use." Gota, Ahmedabad, Gujarat: The Gujarat Institute of Area Planning working paper 18.
Jodha, N.S. 1985a. "Market forces and erosion of common property resources." In *Agricultural Markets in the Semi-arid Tropics*: Proceedings of an International Workshop, October 24–28, 1983. Patancheru (AP), India: International Crops Research Institute for Semi-Arid Tropics.
———. 1985b. "Population growth and the decline of common property resources in Rajasthan, India." *Population and Development Review* 11(2).
———. 1986. "Common property resources and rural poor in dry regions of India." *Economic and Political Weekly* 21(27).
———. 1990a. "Depletion of common property resources in India: micro-level evidence." In G. McNicoll and M. Cain, eds., *Rural Development and Population: Institutions and Policy*. New York: Oxford University Press. Also supplement to vol. 15, Population and Development Review.
———. 1990b. "Drought management: the farmer's strategies and their policy implications." London: International Institute for Environment and Development, Dryland Network Programme issues paper no. 21.

_____. 1990c. "Rural common property resources: contributions and crisis." New Delhi: Society for Promotion of Wasteland Development, Foundation Day Lecture.

_____. 1991. "Sustainable agriculture in fragile resource zones: technological imperatives." *Economic and Political Weekly (Review of Agriculture)* 26(13).

_____. 1992. "Rural common property resources: the missing dimension of development strategies." Washington: World Bank discussion paper 169.

_____. 1993. "Indicators of unsustainability." Kathmandu, Nepal: International Centre for Integrated Mountain Development, MFS discussion paper no. 35.

Magrath, W.B. 1986. "The challenge of the commons: non-exclusive resources and economic development—theoretical issues." Washington: World Resource Institute working paper.

Mishra, P.R., and M. Sarin. 1987. "Sukhomajri-Nada: a new model of eco-development." *Bombay: Business India* (November 16–29).

Ostrom, E. 1988. "Institutional arrangements and the commons dilemma." In E. Ostrom, D. Feeny, and H. Picht, eds., *Rethinking Institutional Analysis and Development*. USA: Institute for Contemporary Studies, International Centre for Economic Growth.

Oza, A. 1989. "Availability of CPR lands at micro-level: case studies of Junagadh programme area of AKRSP (India)." In *Status of Common Property Land Resources in Gujarat and Problems of their Development*. Ahmedabad, Gujarat: Gujarat Institute of Area Planning.

Shah, T. 1987. "Profile of collective action on common property: community fodder farm in Kheda District." Anand, Gujarat: Institute of Rural Management.

Shankarnaryan, K.A. and J.C. Kalla. 1985. "Management systems for natural vegetation." Jodhpur, Rajasthan: Central Arid Zone Research Institute.

Walker, T.S. and J.G. Ryan. 1990. *Village and Household Economies in India's Semi-arid Tropics*. Baltimore: The John Hopkins University Press.

CHAPTER 11

Trees as a Source of Agricultural Sustainability: Agroforestry in China

Runsheng Yin and William F. Hyde[1]

China's agricultural sector grew at the remarkable annual rate of 7.7 percent following the initiation of economic reforms in 1978. Its growth has continued at a rate exceeding four percent since 1984 (China Agricultural Yearbooks 1990-1992). The transition from the collective regime under the People's Commune System was the primary source of growth, but the introduction of market pricing in place of government procurement prices also had an impact (McMillan et al. 1989, Lin 1992, Wen 1993). China's agricultural regions in general, and particularly the northern plains region, experienced a substantial upsurge in agroforestry plantings in roughly the same period.[2] Forest cover in the northern plains increased from about five percent in 1977 to eleven percent in 1988, thereby greatly easing the local construction timber and fuelwood situations, and substantially improving the environment for agriculture (Yong 1989, Zhong et al. 1991). Table 11.1 shows that by 1988 the region possessed 14.66 million hectares of forest shelterbelts, 4.5 million hectares of farmland intercropped with trees, and 1.7 million hectares of small woodlots, along with more than 5.7 billion trees planted around the "Four Sides" of villages, farm houses, roads and waterways (Ministry of Forestry 1988).

The experience of the northern plains provides an unusual opportunity to inquire about the impact of tree cover, or agroforestry, on agricultural growth. Did agroforestry have a positive effect on agricultural productivity in the northern plains, or did the newly planted trees have value only for construction timber and fuelwood? The arguments are well-established that trees in shelterbelts, in alleys intercropped with agricultural crops, and in small woodlots protect against erosion, improve the microclimate, and help restore soil nutrients and soil structure. China's farmers are aware of these contributions of trees to the agricultural environment. Indeed, China probably leads the world in the number of scientists and public agencies examining the role of trees in controlling erosion, and China's government expends considerable effort to transfer the insights of erosion control and other agroforestry research into local application. Nevertheless, the evidence from China and elsewhere on the contributions of trees to the agricultural environment, tends to be specific to research sites and it tends to focus on ecological and physical effects. The economic evidence of widespread positive effects of trees on agricultural productivity is sparse.[3] For China, the broader economic evidence is restricted to Huang and Rozelle's (1995) counter-evidence that environmental stress caused by deforestation has a negative impact on agriculture.

Table 11.1. Forest Resources in Select Areas of Northern China

Province	Area				Standing timber volume[3]	Percent of land covered
	Shelter belts[1]	Inter-cropping[1]	Wood-lots[1]	Four sides[2]		
Henan	2200	1987	223.3	1.13	34.24	12.1
Shandong	3800	1335	400.0	1.28	35.05	7.4
Jiangsu	2000	300	482.0	1.20	25.00	8.2
Anhui	1434	40	276.7	1.15	19.14	13.9
Hebei	3286	47	208.7	0.88	29.00	9.7
Shanxi	976	700	108.0	0.04	17.00	15.0
Shaanxi	967	90	300.0	0.05	7.96	7.2
Total	14663	4499	1698.7	5.73	167.39	

Source: Afforestation Bureau (Ministry of Forestry, 1988)
[1] Measured in 1,000 hectares. Shelterbelts are another term for windbreaks. Intercropping refers to agricultural and tree crops grown intermixed in the same field. Woodlots, of course, refer to small forested areas used for fuelwood, construction wood, and perhaps other forest products.
[2] Measured in billions of trees. "Four sides" are trees planted around villages and farmhouses, and along roads and waterways.
[3] Measured in cubic meters. Excludes the extensive reaches of natural forests. These are outside any usual conception of agroforestry.

The primary motivation for this chapter is to establish the contribution of agroforestry with an example from agricultural productivity in China's northern plains. Is agroforestry's contribution positive? Is it large? Can agroforestry contribute significantly to regional development? We will examine these questions, beginning with a brief historical account of agroforestry in the northern plains, and then turning to an econometric estimation of its contribution to agricultural productivity during the period of remarkable growth since 1978. An aggregate production function in which agroforestry is a production shifter will serve to test the hypothesis that agroforestry does contribute to agricultural productivity. The final section of the chapter reflects on the broader implications of our results.

A Description of Agroforestry in Northern China

Although China has a rich tradition in agroforestry extending over hundreds of years, its current experience is a matter of the past twenty years (Zhu et al. 1991). In the northern plains, few trees, let alone forests, survived the wreckage of war and exploitation that preceded the founding of the People's Republic in 1949. Average agricultural production was approximately 800 kg. of grain per ha. and many hectares of farmland were abandoned due to desertification and salinization caused by flooding, changes in the course of the Yellow River, and other environmental destruction (Henan Yearbook 1986, Shandong Yearbook 1988). Living conditions were abysmal. Even today, approximately 243 million people live on a northern plains land base of 464,000 square kilometers of which only 26 million hectares is farmland, or about 0.08 ha. per capita (Zhong et al. 1991). (This is not atypical of the per capita land base in Southeast Asia. It is on the low end of agricultural experience worldwide.) Furthermore, the base of arable land has been declining at an annual rate of 0.8 percent while the region's population has been growing at 1.5 percent (China Agriculture Yearbooks 1985-1989).

The government began directing attention to land rehabilitation in the early 1950s. Its initial focus was on various environmental engineering and biological measures, including afforestation projects for windbreaks, for fixing sand, and to control salinization. The "Four Sides" tree-planting program began somewhat later. The very limited new forest resources were rapidly depleted, however, after the collectivization and industrialization movements beginning in 1958. Collectivization raised doubts about ownership security for the forest resource, and backyard furnaces for steel production consumed most of the remaining wood for fuel (China Forestry Yearbook 1986).

The economic adjustments of the early 1960s promoted multi-purpose afforestation activities like shelterbelts and intercropping for the purposes of protecting the environment and improving agricultural productivity. Unfortunately, the Cultural Revolution and the political struggles that dominated Chinese society for the next decade restricted any real opportunity for environmental conservation and economic growth. Both agricultural and forest production stagnated. Per capita annual income in the communes was only 62.8 yuan in 1976 (about US $25), more than one third of the rural households were in debt, and many households lived with insufficient fuel, clothing, and housing. Lu (1986) observed that about 100 million people suffered from food shortages.

Following the Cultural Revolution, economic development became the policy priority. Market mechanisms for allocating resources began replacing quotas and price controls. In addition, a contract-based household land management arrangement known as the household responsibility system (HRS) replaced collective production teams as the basic organization for rural production (Lin 1992, Yin 1994). The HRS significantly improved the incentive structure by better connecting effort with reward, and agricultural production has expanded at an unprecedented pace ever since. Grain yields in the northern plains now surpass 6,000 kg/ha/yr (China Agricultural Yearbook 1988) and this region is among the most productive agricultural regions in the world.

China's agriculture has witnessed phenomenal technological adjustments in the past several decades, including the expansion of irrigation, the application of chemical fertilizers and pesticides, the introduction of improved seeds, and the use of plastic sheeting. These technologies have their limitations, however, and several problems have emerged, including a decline in the upper layers of groundwater, a decrease in soil organic content, increasing crop vulnerability to pests and diseases, and extensive crop susceptibility to desiccating early summer winds. Maintaining a sound environment conducive to steady increases in productivity appears to be beyond the reach of modern techniques alone.

An increasing, and properly distributed, forest cover could be an important complementary technology for sustaining agricultural production. Trees can reduce wind velocity and modify solar radiation, thereby regulating air and soil temperature and increasing field moisture. Trees also alter soil nutrition and contain erosion. To be sure, trees do compete with agricultural crops for light and water, and there are limits to the density of forest cover before crop production begins to decline. Nevertheless, China's northern plains was largely deforested by 1970 and the overall potential for tree cover to contribute to agricultural production was probably positive.

Conceptual Method and Data

We can test the hypothesis that the impact of agroforestry on agricultural production was positive by examining an agricultural production function with a term describing the environmental services provided by trees and forests as a production shifter, and with empirical evidence for the period from 1977 to 1990. This approach is equivalent to identifying agroforestry as a technical innovation from the perspective of land management (Hayami and Ruttan 1985). It builds beyond the site specific scientific research evidence that tree cover can complement agricultural productivity by establishing the actual relationship between forest-based environmental services and agricultural productivity over a broad region subject to all the market forces and growing conditions that research trials attempt to control.

Confirming evidence would imply that farmers recognized the potential gains in agricultural and, therefore, that they recognized the overall productivity gains from growing crops and trees together under the strict constraint of available land. It would imply that farmers made a break with the experience of agricultural monocultures in order to internalize the positive externalities contributed by forests to the land management system. If the estimated parameter for forest-based environmental services is positive and significant, then we can infer that agroforestry was an important complement to agricultural productivity, and that the conventional specification of agricultural technology which ignores the environmental services derived from trees and forests is incomplete.

Our basic function takes the form

$$\ln Y_{it} = \alpha + \beta_1 \ln A_{it} + \beta_2 \ln L_{it} + \beta_3 \ln K_{it} + \beta_4 \ln F_{it}$$
$$+ \gamma_1 (HRS)_{it} + \gamma_2 (CM)_{it-1} + \gamma_3 (ES)_{it} + \gamma_4 T \qquad (11.1)$$
$$+ \sum_{i=2}^{5} \delta_i D_i + \epsilon_{it}$$

where Y is the aggregate production, and the terms in logs are the usual inputs to agricultural production: A is land, L is labor, K is durable capital, and F is less durable capital inputs like fertilizer. α is a constant, the βs, γs, and δs are parameters for estimation, the subscripts i and t refer to regional distinctions (prefectures) within our sample population (a province) and the data year, respectively, and the ϵ_{it} are independently distributed error terms with zero means.

The second and third lines of eq. (11.1) describe our production shifters. Two institutional reforms have had tremendous effect on China's recent agricultural performance, the tenurial shift from collectives to the household responsibility system and market reform. The proportion of all production teams converted to HRS is our measure of the tenure reform. CM is an index of the periodic degree of market reform which gradually made agriculture a relatively more attractive economic activity.

It duplicates Lin's (1992) use of the ratio of crop prices to manufactured input prices, lagged one year to account for adaptive expectations for his similar model of China's agricultural production during this period. CM is not a price, and it does not move with either crop or manufactured input prices. It moves with their ratio, getting larger as farmers obtain greater access to the market and, therefore, the average prices received for their crops get relatively larger. ES is our measure of the environmental services provided by agroforestry activities. The D are cross-sectional dummies (one for each of five prefecture). They eliminate the effects of localized variations in factors like soil quality and the extent of irrigation. Finally, a time trend T captures the effects of other unidentified periodic production shifters.

Data limitations restrict us to the Cobb-Douglas form. China's generally unreliable input price data prevent use of a pure cost function, and the limited number of observations available to us restricts us from using a more flexible CES form. These limitations are not a problem. Robust Cobb-Douglas results should be sufficient to test our agroforestry hypothesis.

Data

Our measure of aggregate agricultural output is the sum of the annual physical products of grain, cotton, and oilseed, times their official prices in 1980. This means the 1980 price is the common denominator for aggregation. These three crops account for more than ninety percent of all regional production. The official prices were the only received prices in our initial period, and official and market prices were still not greatly variant by the final period of our analysis. Lin (1992) also used an official price to obtain a measure of aggregate output.

Our measures of agricultural inputs are total sown area for A, total farm labor force for L, and agricultural capital K and chemical fertilizer F. Sown area is a better measure than cultivated area because sown area captures the effect of multiple cropping. The farm labor force is the total rural labor force minus those laborers engaged in the non-agricultural sectors. Our measure of agricultural capital is the total horse power of farm machinery, plus draft animals counted at 0.7 hp per animal. This may be an overestimate because many farm machines are employed in other activities, and we can examine draft animals alone as an alternate measure of capital. Our measure of chemical fertilizer is simply the gross weight of fertilizer consumed. Irrigation is another important capital input, but independent irrigation data are unavailable. Irrigation machinery, however, is part of farm machinery in our measure of capital. Therefore, the coefficient for capital also reflects the impact of irrigation.

Refinements in the Agroforestry Impact

Tree cover (in hectares) is a reasonable proxy for the agricultural-environmental services provided by trees [ES in eq. (11.1)] as long as the trees are well-established and well-distributed in the immediate areas of agricultural production itself. This proxy and our analysis might be refined further by introducing separate agroforestry production shifters for trees intercropped with agricultural products, and for trees in shelterbelts. In each case our proxy is the ratio of area in that agroforestry activity to total farm area.

We identified five potential contributions of trees to the agricultural environment: i) reducing wind velocity and wind erosion, ii) controlling sheet and rill erosion, iii) mediating solar radiation and regulating soil and air temperature, iv) increasing field moisture, and v) adding soil nutrients. We also identified agricultural problems in the northern plains with desiccating winds, decreasing soil organic matter and declines in the upper layers of groundwater.

Desiccating winds have been a massive problem in other regions of China and shelterbelts are specifically designed control them. We might expect Chinese farmers in this region to be particularly alert to their potential. Shelterbelts tend to spread their positive effects among multiple adjacent farms. Therefore, investments in shelterbelts are often entered upon jointly as a communal activity of many local farm households.

Intercropping is mixed agriculture and tree crops in the same field. In this case, trees are not planted densely enough to form good wind barriers and their effect on wind erosion is minimal. Intercropping tends to add the latter four of the environmental contributions of trees fairly uniformly over an entire field. One field, therefore one household, obtains most of the benefits, and intercropping is an investment of individual household choice. In the northern plains, intercropping must be particularly important for its contributions to soil organic matter, and to soil temperature and field moisture. Separating our agroforestry production shifter into two independent variables will allow us to reflect on the production gains from either shelterbelts or intercropping and the wind, organic matter, temperature and moisture effects associated with each.

The objectives of woodlots and Four Sides trees are altogether different and table 11.1 shows that the land area they cover is altogether much smaller than the area covered by shelterbelts and intercropped products. Woodlots (which are small forest plantations) are primarily sources of construction timber and fuelwood although they may help control wind erosion and some are designed to protect river banks and control floods. "Four Sides" trees are planted improve the ambient environment: for example in areas of heavy human activity like the immediate vicinity of homesites. Therefore, the objectives of woodlots and Four Sides trees are not so clearly complementary with agriculture. Indeed, local forestry agencies recognize the differences and compile data on woodlots and Four sides trees in a different form from the agencies' data for shelterbelts and intercropping. All of these are reasons to anticipate that woodlots and Four Sides trees will not have significant impacts on agricultural production and to exclude them from our analysis.

One further consideration is important: the lag between initial seedling establishment and any effect of forest cover on the environment. This lag may be short for both intercropped trees and shelterbelts because farmers in the northern plains plant fast growing trees and because even very young seedlings can crowd competing vegetation and modify soil erosion. The positive effects, especially for shelterbelts, continue accumulating over a long period as their wind-reducing effect only reaches its peak with mature trees. Therefore, we can anticipate that the impacts of young trees probably accumulate rapidly, but the full impacts of shelterbelts may increase well beyond the 13-year period of our data. We examined various possible lags preceding the initial measurable effects, and found the best fit for lags of one year.[4]

Finally, our data are from five representative prefectures in Shandong province for the thirteen year period from 1977 to 1990—which means the data set contains 65 observations. Shandong is the largest and most diverse province in the northern plains region. It is geographically central to and representative of the region. Shandong's Forest Service provided the data for intercropped and shelterbelt areas from their forest surveys of 1977, 1982 and 1988. The forest survey data for intermediate years are interpolations. The provincial Statistics Bureau assisted in compilation of the agricultural data. Table 11.2 summarizes (for the entire province) the full data set in index form, where 1980 is the index year.

Empirical Results

Table 11.3 shows our regression results. We corrected for heteroscedasticity due to different prefecture sizes by dividing all input and output variables by the area of cultivated land in the prefecture in 1980. Otherwise, these are fixed effects models (with a time trend). We detected autocorrelation in the error terms, but correcting for autocorrelation produced little improvement. The first two columns of results report the aggregate production function with the two different definitions of capital. The third column eliminates the physical inputs and focuses on the production shifters as a means of confirming the agroforestry findings from the revenue functions. In general, the statistical fit indicated by the R^2 is satisfying for all three regressions, the coefficients on the agroforestry variables are positive, and the coefficient on intercropping is always significant. The coefficients on most of the remaining variables satisfy expectations, and many of them are statistically significant.

The coefficient on tenure reform is large and significant, while the coefficient on market reform is small and insignificant. These results are consistent with the findings of McMillan et al. (1989) and Wen (1993) for China in general. They provide specific confirmation for Shandong and the northern plains that conversion to the HRS (tenure reform) is the primary source of growth in agricultural productivity. The small market reform effect is not surprising. The data in table 11.2 show that the changes in this index

Table 11.2. Indices of All Variables, 1977–1990

Year	Output	Land	Labor	Capital	Fertilizer	HRS	Market Reform (CM)	Agro-forestry
1978	87.18	101.34	91.88	84.02	71.58	0.00	79.62	17.68
1979	90.12	100.60	96.47	93.46	81.56	0.01	82.05	19.74
1980	100.00	100.00	100.00	100.00	100.00	0.12	100.00	22.78
1981	120.82	97.89	102.80	109.87	119.73	0.40	99.81	25.75
1982	148.31	97.39	103.32	121.73	142.72	0.67	105.61	28.96
1983	175.57	99.84	103.96	145.43	167.60	0.83	104.91	32.25
1984	199.92	102.84	105.08	158.53	171.10	0.95	105.21	35.01
1985	179.99	105.16	104.29	177.81	179.30	0.99	102.29	38.27
1986	180.72	107.25	104.40	196.65	177.04	0.99	96.36	42.10
1987	195.76	106.32	105.38	216.05	176.92	0.99	103.84	46.14
1988	182.83	106.60	107.61	240.89	194.30	0.99	106.99	46.88
1989	192.31	105.72	109.44	245.22	201.98	0.99	114.50	47.29
1990	195.33	99.15	104.16	248.75	215.55	0.99	153.54	47.06

Note: All variables defined in text. Agroforestry is the index for combined shelterbelt and intercropped land shares. The agroforestry index excludes the smaller areas in woodlots and Four-Sides, areas which are not planted for their impacts on agricultural productivity.

were too small in most years to build a case for a significant contribution to agricultural revenues.

The coefficients on the two measures of capital are not significantly different from each other, and the choice between them alters neither our agroforestry results nor the overall statistical reliability of the regressions. This result probably confirms the observation that farmers with more machinery also tend to possess more draft animals, an observation that would be reasonable if draft animals were the major source of field traction and equipment like small tractors was used for other purposes. Finally, the coefficient on fertilizer was also positive and significant.

The coefficients on land and labor are insignificantly negative, and the sum of all coefficients on productive factors is less than one. This would indicate that the production process is characterized by a) little contribution from land and labor and b) decreasing returns to scale. It implies that productivity increases become more and more difficult to obtain over time. This would raise concerns about the outlook for regional agriculture and heighten the importance of exogenous factors like agroforestry that can improve productivity. Since the northern plains is a major contributor to national production and a leader in the recent history of China's overall expansion in agricultural productivity, these findings would also raise questions about the sustainability of national agricultural development without improvements in the agricultural environment. We must examine our land and labor data further, however, before we place confidence in these findings.

Table 11.2 reveals relatively little variation in the land variable data. In fact, land area declined slightly while output expanded rapidly. Therefore, it is not surprising that land is statistically uncorrelated with output. We would conclude that the land variable in our data set has little predictive ability, and the land coefficient in our regressions has little predictive implication.

Labor is a different question. The northern plains population has been growing at a 1.5 percent annual rate—even while the agricultural land base has been declining. Table 11.2 shows that the farm labor force increased in Shandong—even while it fell by approximately one-half in China as a whole! Therefore, our estimated labor coefficient is less surprising.

Nevertheless, we might examine alternative labor data to confirm this conclusion. The State Price Administration Bureau (SPAB) collects province-wide farm labor data for specific crops, and also data on material costs (fertilizer, seeds, irrigation). SPAB's labor data show a decline in farm labor for Shandong that is more consistent with the national experience. The SPAB data do not include any measures of farm machinery or draft animal costs. Furthermore, their province-wide reporting scale precludes the assessment of cross-sectional variation. Nevertheless, we applied the SPAB labor and material data in several different combinations with our data in order to satisfy

Table 11.3. Coefficient Estimates for Agricultural Production

Variable	Specification with full capital	Specification with with draft animals as the only capital	Focus on production shift variables
Intercept	0.361 (0.660)	2.618* (3.829)	2.456* (13.333)
Agroforestry Shelterbelt	0.005 (0.767)	0.001 (0.206)	0.010 (1.204)
Intercropping	0.007* (2.519)	0.006* (2.041)	0.007* (2.033)
Tenure reform (HRS)	0.361* (4.225)	0.551* (6.368)	0.663* (7.909)
Market reform (CM)	0.0002 (0.336)	−0.0001 (−0.277)	0.0001 (0.083)
Capital	0.536* (4.588)	0.346* (5.642)	
Fertilizer	0.435* (4.212)	0.311* (2.966)	
Land	−0.382 (−1.502)	−0.043 (−0.195)	
Labor	−0.064 (−0.189)	−0.172 (−0.549)	
Time trend	−0.065* (−4.525)	−0.035* (−3.405)	
Prefecture dummies	−0.250* (−2.983)	−0.510* (−6.223)	−0.467* (−7.595)
	−0.046 (−0.277)	−0.290 (−1.884)	−0.048 (−0.397)
	0.214* (2.110)	−0.082 (−0.979)	−0.172 (−1.624)
	−0.118 (−1.048)	−0.508* (−4.724)	−0.625* (−5.279)
degrees of freedom	51	51	55
R^2	0.958	0.964	0.914

Expressions in parentheses are asymptotic t statistics. * indicates significance at the 0.01 level.

any concern about our labor and land coefficients. In all cases the labor coefficient became positive, but remained insignificant. The land coefficient generally declined, the sum of coefficients on productive inputs remained less than one, and the key agroforestry coefficients remained essentially unchanged from those reported in table 11.3.[5] This should lend conviction to the observations that our agroforestry findings are robust and that their implications for long-term growth in agricultural productivity are important.

Finally, our time trend is negative and significant. This finding remains when we substitute the SPAB data. It differs with Lin's (1992) finding of a positive trend for all of Chinese agriculture. Our different observation may only indicate that the most important regional technological changes were embedded as quality improvements in capital and fertilizer. [This would be consistent with Stone's (1988) discussion for all of China.] It could also be due to the sharp reduction in government investment in the northern plains' agricultural infrastructure, including poor maintenance and inadequate expansion of irrigation and other public production facilities, and limited accessibility of extension services (Feder et al. 1992). The irrigation infrastructure is greater in the northern plains than in other regions of China, and the decline in its budget was notable.[6]

Our third specification adds conviction to our agroforestry findings. This third specification eliminates the usual agricultural inputs and the trend variable in order to focus on the production shifters. The intercropping coefficient is constant at approximately 0.006 and significant, regardless of the data or model specification. The shelterbelt coefficient is positive but (insignificant and) probably smaller, regardless of the data or model specification. A separate regression shows that the combined effects of intercropping and shelterbelts are in the neighborhood of 0.08 and significant. These results suggest that the soil and air temperature and moisture effects and the soil nutrient effects of intercropped trees are more important to Shandong farmers than the initial wind velocity and wind erosion control effects of shelterbelts. Of course, as the trees mature beyond the 13-year period of our data, the wind controlling effects of shelterbelts will increase. In addition, we have the confirming evidence that government experiment stations have explicitly worked with one species, Paulownia, for this purpose, and that Paulownia shelterbelts in Shandong increased from 113,000 hectares in 1977 to 775,600 hectares in 1988 (Shandong Forest Bureau 1988).

We can conclude that agroforestry in general, and intercropping in particular, had a complementary impact on agricultural production in Shandong, and probably throughout the northern plains. Moreover, agroforestry's impact may increase as the effects of maturing shelterbelts accumulate in the years subsequent to our data. We expect that as the contributions of current institutions and farming systems run their course, farmers will find strong incentives to search for complements like agroforestry that will enable agricultural productivity increases, and they will continue to look for the optimal levels and best applications of agroforestry for enhancing the agricultural environment.

Factor Shares

Decomposing the growth in agricultural into its factor shares will allow us to reflect more closely on agroforestry's contribution. The procedure is to: (1) Multiply the coefficients of the revenue shift variables by 100 (since they are in a semilog form). The coefficients on input variables remain unchanged. (2) Calculate the average change in each variable for the time period in question. (3) And multiply the results from the first two steps for each productive factor. Table 11.2 shows that the tenure reform associated with HRS was essentially complete by 1984. Therefore, we separated this decomposition into two periods, 1977–1984 and 1985–1989. Columns 4 and 6 of table 11.4 report the factor shares for the second regression in table 11.3. The first number in these columns is the absolute share of agricultural growth while the parenthetical expression underneath it is the percentage share.

The conversion to HRS explains over half of the total growth in the first period. This observation is roughly comparable to Lin's (1992) estimate of 47 percent of agricultural growth nationwide due to HRS. McMillan et al (1989) and Fan (1990) make similar observations. Aside from the change in tenure; capital, fertilizer, and forest-related agricultural services were the major contributors to productivity growth. Agroforestry, including the impacts of both shelterbelts and intercropping, contributed moderately increasing absolute shares over the two periods (4.97 and 5.39 in 1977–1984 and 1985–1989, respectively). Its proportional share was only 5.42 percent during the earlier period when changes in farming institutions induced the most rapid growth, and when many agroforestry tree plantings were younger. Agroforestry's proportional share increased to 19.4 percent in the second period as the trees and their impacts matured. This second period result supports the casual observations of Zhong et al. (1991) and Zhu et al. (1991) of 10–15 percent increases in crop yields.[7] Indeed, their estimates may have been conservative for the northern plains. Most of the contribution of agroforestry in Shandong was due to intercropping, which by itself accounted for 4.24 percent of the growth in agricultural productivity between 1977 and 1984, and 17.72 percent between 1985 and 1989. The larger effects of shelterbelts should become apparent later as most shelterbelts will only mature after 1990.

Table 11.4. Accounting for Productivity Growth

	Regres'n coeffic't	1977–1984		1985–1989	
		Avg chg in variable	Contrib'n to growth[1]	Avg chg in variable	Contrib'n to growth[1]
Inputs					
Capital	0.346	74.51	25.78 (28.05)	86.69	29.99 (107.66)
Fertilizer	0.311	99.52	30.95 (33.67)	30.88	9.60 (34.47)
Land	-0.043	1.50	-0.065 (-0.07)	2.88	-0.12 (-0.44)
Labor	-0.172	13.20	-2.27 (-2.47)	4.36	-0.75 (-2.69)
Shifters					
Tenure reform	0.551	0.99	54.55 (59.34)	0.00	0.00 (0.00)
Market reform	-0.0001	25.59	-0.20 (-0.28)	9.29	0.09 (0.32)
Trend	-0.035	7.00	-0.25 (-0.27)	5.00	-17.50 (-62.81)
Agroforestry					
Shelterbelts	0.001	10.84	1.08 (1.18)	4.54	0.45 (1.63)
Intercropping	0.006	6.49	3.89 (4.24)	8.23	4.94 (17.72)
Residual			2.77 (2.99)		1.16 (4.15)
Total growth			91.92 (100.00)		27.86 (100.00)

[1] Percentage contributions in parentheses.

Concluding Remarks

The nearly complete removal of forest cover from China's northern plains before the renewal of household incentives and renewed planting of trees in the late 1970s provides an unusual broad-scale opportunity to examine the with- and without impact of forest-related environmental services on long-term agricultural productivity. Our evidence for Shandong province in this region shows that agroforestry activities, and intercropping in particular, rapidly paid off with greater agricultural production. Furthermore, it is probably true that the contribution of agroforestry activities has not reached its limit for Shandong, or probably for the entire northern plains region.

The relevance of this finding goes beyond China's agriculture and the local complementarity of trees with agriculture. This finding also provides an insight to the much broader question of how world agricultural production can be sustained in the long-run—after the possible exhaustion of the contributions of current institutions and farming systems. If the opportunities for expanding the agricultural land base and for multiple cropping are small, and if ever-more labor must be squeezed into the agricultural sector, then what can contribute to agricultural growth? Part of the answer must lie in improvement of the agricultural environment, and trees—along with R&D, education and extension, and investments in infrastructure—may be a part of this improvement.

The Shandong experience is part of a mounting body of evidence that trees are increasingly economically scarce in many areas around the world, that the values of trees are increasingly internalized in local decisions, and that these values are less and less often restricted to residual resources growing at the external margin of productive land.[8] As trees become scarcer with continuing deforestation, and as agricultural productivity slows in regions like China's northern plains, then farmers will respond. Public research and development agencies that are alert to the economic indicators of forest scarcity, and also to the regions where private agroforestry responses are promising, will be able to provide significant support for agricultural and environmental development and sustainability.

NOTES

1. University of Georgia, Athens, Ga., and Centre for International Forestry Research, Bogor, Indonesia. S. Rozelle and an unidentified reviewer provided important critique and useful insight.

2. The northern plains covers roughly Hebei, Liaoning, Henan, Shandong, Beijing, Tianjin, the northern parts of Anhui and Jiangsu, and the central part of Shaanxi.

3. Nair (1993) is the classic reference for general agroforestry practices. Nair makes the same point that economic evidence is sparse, while the evidence from biological field trials is more thorough, if still incomplete.

4. This lag should remove any doubt about simultaneity between reforms and agroforestry production. Agricultural reforms preceded any successful agroforestry impact by a period at least as long as the one year lag. (Reforms for the wood market came even later and would have had a longer lagged effect on agricultural productivity.)

5. Regressions available from the authors.

6. This suggests that our trend variable may capture both some of the usual positive effect of technical change and a negative effect of declining infrastructure, with the latter dominating in our variable. Antle (1983), for a cross-section of developed and developing countries, showed that a measure of infrastructure can be highly important.

7. A separate regression shows an even higher proportional share for the combined effects of intercropping and shelterbelts taken together.

8. Hyde et al. (1996) review the empirical economics evidence, including more than 60 other citations. They also place this evidence in the context of economic development and global deforestation.

REFERENCES

Antle, J. 1983. "Infrastructure and aggregate agricultural productivity: International averages." *Economic Development and Cultural Change.* 31(3):609–19.

China Agricultural Yearbooks. 1985–1992. Beijing: China Agricultural Press.

China Forestry Yearbook. 1986–1988. Beijing: China Forestry Press.

Fan, S. G. 1990. "Effects of institutional reform and technological change on production growth in Chinese agriculture." *American Journal of Agricultural Economics* 73(2):266–75.

Feder, G., L. J. Lau, Y. J. Lin, and X. P. Luo. 1992. "The determinants of farm investment and residential construction in post-reform China." *Economic Development and Cultural Change* 40(3):287–312.

Hayami, Y. and V. W. Ruttan. 1985. *Agricultural Development: An International Perspective.* Baltimore, MA: The Johns Hopkins University Press.

Henan Yearbook. 1986. People's Press of Henan.

Huang, J. K., and S. Rozelle. 1995. "Environmental stress and grain yields in China." *American Journal of Agricultural Economics* 77(4):853–964.

Hyde, W. F., G. S. Amacher, and W. Magrath. 1996. "Deforestation, scarce forest resources, and forest land use: theory, empirical evidence, and policy implications." *World Bank Research Observer* 11(3):223–48.

Lin, Y. J. 1992. "Rural reforms and agricultural growth in China." *American Economic Review* 81:34–51.

Lu, X. Y. 1986. *Research on Responsibility Systems.* Shanghai: People's Press of Shanghai.

McMillan, J., J. Whalley, and L. J. Zhu. 1989. "The impact of China's economic reforms on agricultural productivity growth." *Journal of Political Economy* 97:781–807.

Ministry of Forestry. 1988. *Resource Bulletin.* Beijing.

Nair, P. K. R. 1993. *An Introduction to Agroforestry.* Dordrecht: Kluwer Academic Publ.

Shandong Forest Bureau. *1988. Forest Statistics in Shandong Province.*

Shandong Yearbook. 1988. People's Press of Shandong.

Stone, B. 1988. "Development in agricultural technology." *China Quarterly* 116:762–822.

Wen, J. G. 1993. "Total factor productivity growth in China's farming sector: 1952–1989." *Economic Development and Cultural Change* 41:1–41.

Yin, R. S. 1994. "China's rural forestry since 1949." *Journal of World Forest Resource Management* 7:73–100.

Yong, W. T. 1989. "Agroforestry development and its implications." *Forestry Problems.*

Zhong, M. G., C. Xian, and Y. M. Li. 1991. "A study of forestry development in the major agricultural areas." In Yong, W.T. ed. *Studies on China's Forestry Development*. Beijing: China's Forestry Press.

Zhu, Z. H., Cai M. T., Wang S. J., and Yiang Y. Y. (eds). 1991. *Agroforestry Systems in China*. Jointly Published by Chinese Academy of Forestry, People's Republic of China, and International Development Research Center, Canada.

Part 7
Conclusions

Our final chapter is a review chapter. Its intention is to review and integrate the material from the previous eleven chapters. As we have prepared the material for this book, others have been examining many of the same topics, and from similar perspectives. This chapter will incorporate observations from their findings as well as our own.

The chapter reviews the expectations for forestry's contribution to rural development—and for its special contributions to the most disadvantaged, to women and the landless users of the forest commons. A growing literature challenges some of these expectations—although this literature could also be used to support a call for sharper definitions of deforestation, improved indicators of the effects of forest resources on the rural poor, and improved design of forest policy interventions. We review this literature, suggest some unifying themes, and identify the critical issues that remain unanswered.

Our fundamental contention is that policy interventions should be analyzed with regard to markets, policies, and institutions. Markets for forest resources generally exist in some form—although they may be thin. Successful forestry projects and policies require careful identification of the target populations and careful estimation of market and market-related effects on the household behavior of these populations. They also require respect for the institutional structures that assure secure rights for scarce forest resources. Social and community forestry, improved stoves, improved strains of multipurpose trees, and even private commercial forest operations can all improve local welfare where scarcity is correctly identified and the appropriate institutions are in place. The increasing number of observations of afforestation from developing countries around the world is evidence that forestry activities do satisfy these conditions in selective important cases. The critical point for policy is to identify the characteristics of these successful cases that are predictive of other cases where new forestry activities can be welfare enhancing.

CHAPTER 12

Social Forestry Reconsidered

William F. Hyde, Gunnar Kohlin, and Gregory S. Amacher[1]

Social forestry generally refers to the range of activities associated with forest products, the rural environment, and subsistence agricultural communities. It often features external development assistance intended to benefit these communities and their environments. Over the past ten years, the prevailing wisdom on social forestry as a welfare enhancing technology has run the full course from optimism, even enthusiasm, to a current attitude of caution, and even scepticism. Meanwhile, a developing body of analytical literature is beginning to provide evidence. This literature consistently supports the optimistic view—but only under carefully selective conditions.

This is a good time to examine the evidence, to identify reasonable hypotheses about the successful uses and limits of social forestry, and to identify the important questions that are yet to be addressed. This is our objective.

Our fundamental contention is that successful social forestry activities must be assessed in terms of their contributions to human welfare. This contrasts with a view that forest cover, therefore forest protection and afforestation, is a useful end in itself. The latter view bolsters policy decisions to halt deforestation or policy objectives to maintain a fixed share of land in forest cover, "one third of the land" in the case of India (Rao and Srivastava 1992) or all land with greater than an eighteen percent slope in the Philippines.

Physical standards are poor measures of human welfare. Welfare, including the welfare gains due to social forestry activities, is revealed by household preferences in the context of local markets, institutions, and policies. Furthermore, we anticipate that those rural households that are affected by social forestry follow fairly standard expectations of behavior with respect to these markets and the policies that affect them. This perspective is consistent with Dewees' (1989) earlier observations for social forestry, with Schultz' (1964) observations for the broader experiences of farm households, and with the general literature of rural development.

Nevertheless, some important behavioral differences distinguish household participation in agriculture from participation in social forestry. The open access condition of the natural forest is the source of these differences. This condition is common to a large component of the world's natural forests. It exists because more secure property rights cost more than some share of the natural forest is worth. Therefore, open access does not indicate a market failure in this case. This open access condition accompanies important gender and wealth distinctions in the classes of resource users and seasonal differences in forest use itself. It supports insights to resource use (by whom, when, and for what purposes) that are different from the agricultural experience where households tend to have more secure rights to their lands and capital. Questions about deforestation, sustainable use, and reforestation each require insights to the use of the open access forest and to the conditions that cause local farm households to shift from reliance on this resource to the secondary forests that are beginning to appear on the farmers' own agricultural lands where land use rights are more secure.

Our paper begins by revisiting the social forestry concept itself and the general expectations many held for its beneficial impacts on rural communities and their environment. These expectations lead us to the new questions, as well as the doubts that practitioners eventually began to raise. Both the original expectations and the new questions induce us to review the empirical evidence. We will review the evidence on i) consumption, ii) production and supply, and iii) investment in new technologies, before submitting a summary characterization of the general forest environment, and then closing with our view of the lessons for policy analysis.

We will find that market incentives identify the most likely households to invest in any new technology only when the return is good, the rights to the technology and its products are secure, the risks are acceptable, and the policy environment is stable and predictable. We think that some, but not all, of the critical market information is well understood, and that there is some comprehension of the impacts of forest policies and other institutions. Nevertheless, we think the factors that explain which regions and households actually do invest are poorly understood, that the opportunities for improved performance by local institutions are largely determined by unexamined opinions, and that there is inadequate recognition of the crucial impact of the overall national or regional policy environment.

Background: Prevailing Wisdom

As the concept of social forestry has become more widely accepted, its definition has become more elusive. It includes local community and local private activities, often by subsistence households. For our purposes, it refers to the production and use of fuelwood, forage and fodder, fruits and nuts, latex, gum, and various other non-timber forest products. It includes domestic uses and local market exchange of construction timbers—but it generally does not include industrial wood production or domestic woodlot production for shipment beyond local markets.

The applications of social forestry have grown beyond its original conception as seedling distribution, planting and technical assistance to incorporate watershed management and the broader class of forest contributions to the natural environment. Its practitioners recognize the soil sustaining properties of trees and good forest management. They also recognize the indirect gains from substitution; in particular the gains from substituting woodfuels for agricultural residues and the gains from introducing improved stoves. When woodfuel replaces dung and straw the latter remain in the fields where they add the structure and nutrients that help sustain depleted soils. Improved stoves decrease the consumption of combustible material and, thereby, save both forests and soils.

Social forestry is viewed all the more favorably because those who benefit from it are often the most disadvantaged: women, and the rural poor, and especially landless users of the forested commons. In many developing economies it is a woman's task to collect the materials provided by the forest, and the forested commons is a resource of last resort for the poorest households. In times of greatest economic stress, it can become a source of both food and marketable products.

In sum, social forestry has always had a local household or community orientation and the effects of activities that increase the occurrence of trees and forests on the local landscape seem to be uniformly positive. The contributions of social forestry to global concerns for reforestation and environmental sustainability are clearly positive as well.

The rural character of forestry means that many market transactions and a substantial amount of consumption from the forest never appear in any country's national accounts. Nevertheless, various estimates have been calculated. They suggest, for example, that hunting generates 20-50 percent of cash income for forest villages in developing countries, and that wood constitutes fourteen percent of overall energy consumption in developing countries in general and nearly fifty percent of energy consumption in Africa (cited in Persson 1998). One summary measure comes from the best attempt to date to incorporate non-market environmental and resource values in the national accounts of the Philippines. Peskin and delos Angeles' (IRG with Edgevale Assoc. 1994) estimate that subsistence household use of the forest is the single largest undervaluation in the Philippine accounts, larger than water or air pollution or soil depletion, and larger than subsistence consumption of agricultural or fisheries products. It is a reasonable hypothesis that subsistence households in other developing countries obtain comparable large relative values from their forests as well.[2]

The institutional budget for forestry is good evidence of a policy commitment that is consistent with these observations. The World Bank's recent annual commitment has been in the neighborhood of US$250 million. Through the Global Environmental Facility the Bank administers another US$20-25 million annually to protect biodiversity that largely occurs in forest environments. Bilateral aid agencies, regional development banks, and local government agencies contribute additional sums. In India, for example, social forestry projects were 25 percent of the national budget for rural development in the late 1980s (Sharma, McGregor, and Blyth 1991). India paid a staggering 35 billion rupees for afforest 13 million hectares (Kajoor 1992). Almost all such funding activities have important links to the local human populations that live in and around the forest.

Finally, the spontaneous and independent tree planting undertaken by local farmer households in many local situations in developing countries around the world is the best evidence of the on-the-ground success of social forestry. Our own experiences include examples in Pakistan, India, Nepal, Indonesia, the Philippines, China, Vietnam, Ethiopia, Kenya and Malawi. Often the local magnitude of the tree planting activity is small but it can accumulate to real importance across broad regions. In Bangladesh, for example, local farm households accounted for 3/4 of all marketed timber and fuelwood in 1980 (Douglas 1982). They probably account for an even larger share now.

Nevertheless, despite the evidence of success, there have been many failures as well, and these failures have generated a healthy scepticism among practitioners. More constructively, evidence of both sterling successes and stunning failures encourage questions of "why some social forestry activities succeed while others fail" or "what distinguishing characteristics of the successes would predict success elsewhere in future introductions of new social forestry technologies."

We might anticipate that failures occur where non-forest consumption and non-forest uses of household labor are more important than forest consumption and production. Surely it is reasonable that some poor households and some communities have more immediate concerns with food than with forestry. This is an income or household budgetary effect and we can anticipate greater likelihood of success where forestry occupies a larger share of the household budgets of income or effort. In addition, many have pointed to the importance of local institutions, and especially to well-established property rights for trees and forest land, as prerequisites for successful long-term forest investments.[3] Finally, Byron (1997) points out that the changes that occur as normal development proceeds can be deterrents to social forestry. For example, the demands for some forest products decline as incomes rise. As the values of forest products rise the pace of substitution away from them also increases. Finally, the expanding commercialization that accompanies development modifies gender roles, inter-household relationships, and the competition for property rights and, thereby, modifies production and consumption and the attraction of specialized social forestry activities.

Can we build on these insights on income effects, property rights, and changes in overall economic conditions to create a broad basis for anticipating the locus of successful social forestry investments? That is our task in this paper.

Responses to Scarcity

Both global concerns for deforestation and local concerns for improved household welfare can be expressed in terms of scarcity and prices. Greater deforestation means increasingly scarce forests; therefore, increasingly scarce forest products. Increased scarcity means higher prices and greater opportunity for welfare enhancement by any of the myriad of social forestry activities that decrease forest consumption or expand forest production.

Godoy (1992) supports this argument with 21 regional examples of farmers in Africa, Asia, and Latin America who responded to high forest product prices by planting trees. Godoy also points out, however, that high prices are not a sufficient condition for tree culture. Our own observation is that rising prices flag a potentially emerging opportunity for social forestry but an opportunity that is realized only when prices rise to a level sufficient to induce substitution, either substitution of alternative sources of forest products or substitution of altogether different non-forest products for the forest products.

The first part of this argument, and Godoy's evidence, encourage us to examine the literature on household price responses in the consumption and production of forest products. The latter part of the argument—that high prices alone are insufficient—encourages us to examine other factors as well as the own-price of forest products, factors like household income and demographic characteristics and factors like the prices of substitutes.

Markets may be thin but they do exist—even for most minor forest products. Moreover, even subsistence households generally participate in some markets.[4] Even when they do not trade in the market for a particular product, they could—because local markets do exist. And households would participate if market prices fell sufficiently to induce them to purchase rather than to expend their own scarce labor collecting the product for their domestic use. When households collect for their own use, they are revealing that the value of the labor they expend in collection is less than the market price of the product but greater than the wage their labor would earn in alternative employment.

This reasoning means that market information on forest products is valid even in subsistence communities, but it is insufficient because the household's consumption and production decisions are inseparable. That is, household production decisions affect consumption, and *vice versa*. An economic modelling approach known as the new household economics accounts for this non-separability. It has been applied widely in agriculture, and more recently in several forestry cases. We will rely on this literature and its evidence from surveys of rural households to sort out the factors explaining household responses to scarcity.

Rural Household Consumption of Forest Products

Table 12.1 summarizes the empirical evidence on rural household responses to market prices for forest products. It features the primary forest products in each region of inquiry, but the preponderance of evidence is on fuelwood because fuelwood is the most widely used forest product for rural households. For example, market fuelwood purchases account for as much as 20 percent of household cash outlays in central Malawi (Hyde and Seve 1993), and market fuelwood production is the largest share of the 38 percent of all household income that comes from the forest in one district of Sri Lanka (Bogahawatte 1997). The FAO estimate for the annual world value of fuelwood is US$42 billion, perhaps 10-15 percent of which is sold in the market (cited in Persson 1998.)

Most observers have found that the household consumption of forest products is price inelastic regardless of the forest product—although consumption is less inelastic in the arid uplands of Ethiopia for example (Mekonnen 1998), than in the cool moist hills of Nepal (Cooke 1996). Cooke noted that household demands are more price inelastic when corrected for seasonal differences in consumption patterns—influenced by seasonal weather and enabled by seasonal labor availability and household storage. Cooke's households are also more wage responsive than price responsive, a condition that raises two possibilities: a) as prices rise, the implicit wage for collecting forest products for the household's domestic use becomes a relatively more important determinant of household consumption, or b) higher wages are indicative of higher incomes and household income or wealth is more important than market prices and a determinant of consumption.[5]

Amacher, Hyde, and Kanel (AHK 1998) drew similar conclusions for fuelwood in Nepal's lower elevation and drier tarai region—although not for the breadth of Nepal's hills—and they agreed with Cooke that wages are a more important factor than prices in the household consumption decision.

Table 12.1. Consumption Responses: Price

Study	Location	Forest Product	Measure	Elasticity
Cooke (1998a)	Nepal's hills	fuelwood	demand shadow price	−0.25*
		forage	demand shadow price	−0.10*
		fodder	demand shadow price	−0.11*
Mekonnen (1998)	Ethiopia	fuelwood	demand shadow price	−0.40*
		dung	demand shadow price	−0.72*
Amacher, Hyde, Kanel (1998)	Nepal's hills	fuelwood	market demand price	−1.47*
	Nepal's tarai	fuelwood	market demand price	−0.21
Amacher, Hyde, Joshee (1993)	Nepal's hills			
	low income	fuelwood	collection time	−0.28*
	high income	fuelwood	collection time	−0.84
Heltberg, Arndt, Sekhar (1998)	Rajasthan, India	fuelwood	collection time	−0.11*

Notes: a) Cooke and AHK found that household consumption was more responsive to (implicit) wages than to prices. Mekonnen found that consumption was at least as wage responsive as price responsive. b) Dung is found in the fields, but also in the same open access lands that are sources for fuelwood, forage, and fodder. c) Collection time is a proxy for labor opportunity cost in AHJ and HAS. * indicates statistical significance at least at the 0.10 level.

Amacher, Hyde, and Joshee (AHJ 1993) and Heltberg, Arndt, and Sekhar (HAS 1998) might take the argument about the relative importance of wages a step further. They found either little price variation (AHJ) or little evidence of market purchase (HAS). Rather than price, they chose to focus on collection time, a measure of the labor opportunity cost and a proxy for the importance of the household wage. Both AHJ and HAS observed inelastic household responses to this measure. Inelastic collection time means that, while households do respond to deforestation and increasing scarcity by decreasing consumption, the consumption effect is small and it is dominated by offsetting increases in collection time. This is not a satisfying finding for those who anticipate that rational household behavior will solve the problems of fuelwood scarcity and deforestation. If the major response to scarcity is an increase in collection time, then increasing scarcity does not induce increased forest protection.

This finding also raises questions about the source of any increase in collection time. Is labor a slack variable, such that households forego little by extending greater effort in fuelwood collection? Or does an increase in collection time mean that less time is available for household chores like agricultural activities, food preparation, and childcare? Many have presumed the latter. We will turn to the empirical literature on the question shortly.

Income Effects

The observation on the relative importance of wage effects was consistent across all of these analyses. This observation, together with Byron's suggestion that forest products may even be inferior goods (consumption declines as incomes rise), encourages us to examine income effects. Negative income elasticities would indicate inferior goods. Small positive income effects would indicate necessary goods which households consume at relatively constant levels regardless of their wealth. Inferior goods and necessary goods typically (not always) consume a small share of household budgets and are of small consequence to many important decisions of household behavior.

The seven analyses that are reported in Table 12.2 suggest this is the case for forest products. The effect of household income on the consumption of forest products is generally small, and some forest products are inferior goods in some economies. Furthermore, two of these analyses suggest that the household's income source does not alter this finding. The household's income source suggests something about its lifestyle, or at least how dependent the household is on its own agricultural land. Mekonnen and AHJ both found that it does not matter whether the household relies on income from its agricultural production, from the hire of its labor, or from remittances. The household's income level has only a small effect on its consumption of the most important forest products.

The income effect is one reason why many households and communities around the world have not been especially receptive to external forestry assistance despite local deforestation and rising prices for forest products. And this argument corresponds well with Godoy's point that high prices are important, but that they are an insufficient incentive for many social forestry activities.

Substitution

To find households with sufficient incentives, we need evidence of substitution away from the products of natural forests or into substitutes from secondary forest plantations. Table 12.3 reviews the evidence for substitution in consumption. One study (Cooke) examined forage and fodder substitution. The remainder feature fuelwood and its substitutes. Commercial fuels like kerosene or LPG seldom penetrate these rural markets. Therefore, the common substitutes for fuelwood are combustible agricultural residues (like straw and dung) and improved stoves, a technological substitute.

Table 12.2. Consumption Responses: Income or Wealth

Study	Location	Forest Product	Measure of income/wealth	Elasticity (or coefficient)
Cooke (1998a)	Nepal's hills	fuelwood	upland land area	(+)
		forage	upland land area	(−)*
		fodder	upland land area	(+)
Mekonnen (1998)	Ethiopia	fuelwood	labor income	0.063*
		fuelwood	non-labor income	0.03
		dung	labor income	−0.02
		dung	non-labor income	−0.02
Amacher, Hyde, Joshee (1993)	Nepal's hills low income	fuelwood	agric. income	−0.31*
		fuelwood	exogenous income	−0.20*
	high income	fuelwood	agric. income	0.0005
		fuelwood	exogenous income	0.002
	low income	agric. residues	agric. income	0.36*
		agric. residues	exogenous income	0.05
	high income	agric. residues	agric. income	−0.0004
		agric. residues	exogenous income	−0.001*
Amacher, Hyde, Kanel (1998)	Nepal's hills	fuelwood	farm size	0.0005
	Nepal's tarai	fuelwood	farm size	0.07
Heltberg, Arndt, Sekhar (1998)	Rajasthan, India	fuelwood	farm size	−0.25*
Shyamsunder, Kramer (1997)	Madagascar	fuelwood	rice production	−0.01*
		palm leaves	rice production	−0.01*
		wood crab	rice production	0.00
Bogahawatte (1997)	Sri Lanka low income	fuelwood	total income	< 1
		med. plants	total income	< 1
		mushrooms	total income	< 1
	high income	fuelwood	total income	< 0
		med. plants	total income	< 0
		fuelwood	total income	< 0

Notes: a) The household's private land area is a proxy for income or wealth for Cooke, AHK, and HAS. Household rice production is a proxy for income for SK. b) Agricultural residues include dung, and also crop residues like straw. Dung is found everywhere that cattle roam, including in the forest. Straw, of course, is not a forest product. c) Cooke did not calculate an elasticity but we can infer elasticities of less than or greater than one from her estimated coefficients and sample size.

Table 12.3. Consumption Response: Substitution

Study	Location	Forest product	Measure of substitution	Elasticity (or coefficient)
Cooke (1998a)	Nepal's hills	forage fodder	fodder shadow price forage shadow price	(−) (−)
Mekonnen (1998)	Ethiopia	fuelwood dung	dung shadow price fuelwood shadow price	−0.3* −0.7*
Amacher, Hyde, Joshee (1993)	Nepal's hills[2] low income high income low income high income	fuelwood fuelwood ag. Residues ag. Residues	impr. stove owner impr. stove owner fuelwood coll'n time fuelwood coll'n time	(−)* (+) (+) (+)
Amacher, Hyde, Kanel (1998)	Nepal's hills mkt. partic. collectors Nepal's tarai mkt. partic. collectors	fuelwood fuelwood fuelwood fuelwood	impr. stove owner impr. stove owner impr. stove owner impr. stove owner	(+) −0.33* (+) −0.26
Heltberg, Arndt, Sekhar (1998)	Rajasthan, India	fuelwood	impr. stove owner	(−)

ªNote: Where elasticities were not calculated, we can infer substitution or complementarity from the signs on the estimated coefficients and the sample sizes. This evidence on substitution must be read carefully because some coefficients refer to prices or price proxies (collection time) while some refer to quantities (wood, stoves). * indicates statistical significance at least at the 0.10 level.

Neither Cooke nor Mekonnen found evidence of substitution. Indeed, both observed complementarity, between forage and fodder as animal food in Nepal (Cooke) or between fuelwood and dung for cooking and heating in Ethiopia (Mekonnen). The latter is an especially interesting finding because the general view holds that dried dung is a primary substitute for fuelwood and that the combustion of dung only decreases its contribution to soil sustainability, long-run agricultural productivity, and nutrition in subsistence agricultural communities. Alternatively, increasing tha availability of fuelwood decreases the negative effects of burning dung. Mekonnen's evidence urges caution on global generalizations of this view.

AHJ did find evidence of substitution in their narrower survey of two districts in Nepal's hills, substitution between fuelwood and agricultural residues, and especially for low income households. They also observed that low income households make the technological substitution of improved stoves for fuelwood.[6]

The real importance of AHJ's observation has to do with target populations. Fuelwood prices were high enough to induce substitution—but only for lower income households. AHK confirmed this point and they sharpened the definition of the target population. In a broader survey of districts across Nepal's hill and tarai regions, AHK observed that as prices rise some households turn from market fuelwood purchases to fuelwood collection to satisfy their own domestic consumption. Households that collect but do not also participate in the market, which are also lower income households, and especially in the hill region, are statistically significant substituters of improved stoves for fuelwood.

Summation

The consumption evidence tells a consistent story. Households do respond to higher forest product prices by decreasing their consumption. However, the household income elasticities for forest products are generally small. This indicates that forest products are generally necessities and also that we should expect only the poorest households to be especially responsive to their scarcity. This is not encouraging for the hypothesis that increasing scarcity will slow rates of deforestation, of for the widespread and general introduction of social forestry activities. Select cases, however, offer more promise. These select cases can be identified by populations that are sufficiently responsive to higher prices that they substitute for their consumption of forest products. In the examples from Nepal, this was not an entire regional population. Rather, it was a poorer subset of the entire population from a region that was relatively more price responsive as a whole. Of course, this is precisely the population that public development projects intend to assist.

Household Production

Household production is more complex. The range of household options is broader, including collection from alternate sources of the product, labor reallocation among the household's subsistence activities (including fuelwood collection), and labor reallocation to wage and income producing activities—which then allow the full range of consumption alternatives. The breadth of production alternatives has led to a more diverse empirical literature, and our generalizations from it must be more speculative.

Market Supply

Table 12.4 summarizes the three insights to household supply known to us. All three observed some degree of price responsiveness, whether the households were supplying their own domestic consumption or the local fuelwood market. The latter, market suppliers, are particularly important and they are easy to overlook. Most surveys of non-timber forest products intend to be random and unbiased. Nevertheless, they tend to observe substantial levels of market demand but only occasional market supply. This means that they overlooked some market-supplying households. Unfortunately, these are probably lower income and landless households, the households of our greatest general concern.

Table 12.4. Production Responses: Supply Prices

Study	Location	Forest product	Measure	Elasticity
Kohlin (1998)	Orissa, India market suppliers	fuelwood equivs.	market price collection shadow wage	—
Amacher Hyde, Kanel (1998)	Nepal's hills market suppliers	fuelwood	market price collection shadow wage	2.99* 6.81* 1.57*
	domestic collect.	fuelwood	market price collection shadow wage	1.33* 0.36*
	Nepal's tarai market suppliers	fuelwood	market price collection shadow wage	0.55 0.71
	domestic collect.	fuelwood	market price collection shadow wage	0.09 0.41
Heltberg, Arndt, Sekhar (1998)	Rajasthan, India market suppliers	fuelwood	market price	

Note: Kohlin's fuelwood equivalents are measures of the combustion equivalent to fuelwood that is obtained from all burnable forest products: leaves, twigs, and larger fuelwood. * indicates statistical significance at least the 0.10 level.

Kohlin's (1998a) evidence from Orissa in India supports this contention. Kohlin found that women from lower caste and lower income households were the most likely suppliers of market fuelwood. They are also more likely to obtain their market fuelwood from the open access natural forest than from their communities' managed village woodlots—although the village woodlots were always closer than the natural forest in Kohlin's region. This seems to confirm the evidence of several (notably Jodha 1999) that open access resources are resources of final resort for the poorest households.

AHK's evidence largely supports Kohlin, although it is less complete in its description of the market suppliers themselves. It is also consistent with AHK's consumption evidence (tables 12.1-12.3) which shows regional differences in fuelwood scarcity in Nepal and shows that, beyond some level of scarcity, rural households do respond to increasing scarcity by altering their consumption behavior.

Finally, HAS examined consumption, not market supply, but they observed that twelve percent of households in India's Rajasthan sold to the market. Market sales represent 26 percent of all collection for these households and these households are (insignificantly) positively price responsive.

Household Production, Land and Labor Inputs, Gender, and the Poor

Table 12.5 summarizes the more extensive evidence on household production itself. Evidence from a range of cases (Nepal, Madagascar, Ethiopia, India, Malawi) supports the contention that the collection of fuelwood and other forest products declines with decreases in the available forest stock and with decreases in the accessibility of the remaining stock. Furthermore, under conditions of sufficient scarcity, private stocks do become a substitute for forest resources on the common lands. For example, in contrasting districts in Nepal's hills AHJ observed that when fuelwood becomes sufficiently scarce on the community's common lands (smaller and less accessible stock and higher prices), households eventually begin growing wood on their own private lands. Nevertheless, all observers agree that as the stock declines or becomes less accessible, households generally spend more time collecting forest products.

Labor allocation to the collection activity has been a critical question because so many forestry and women-in-development projects presume a unique cultural role for women. Who collects from the forest? Is collection the special responsibility of women, or of women and children? And does an increase in time spent collecting forest products detract from other household responsibilities, in particular the food preparation and agricultural production responsibilities that are crucial to survival in subsistence rural communities?

In fact, Shyamsunder and Kramer (1997), Mekonnen, and AHJ all found that labor allocation varies between genders. Collection is not the domain of women alone, whether in Madagascar, Ethiopia, or Nepal, and on some occasions men are the primary collectors.[7]

Table 12.5. Production Responses: Land and Labor Inputs

Study	Location	Forest Product	Land and Labor Imputs
Amacher, Hyde, Joshee (1993)	Nepal's hills	fuelwood	If more plentiful supply: collect from commons; women collect, men do not, children negative collectors.
	Nepal's hills	fuelwood	If more limited supply: rely more on own lands, capital inputs significant, men and women collect but men collect more.
Amacher, Hyde, Kanel (1998)	Nepal's hills	fuelwood	Level of resource stock and resource access significant.
	Nepal's tarai	fuelwood	Levels of stock and access significant, some indication of supply from own land and of increased supply from lands of wealthier households
Shyamsunder, Kramer (1996)	Madagascar	fuelwood palm leaves wood crab	Collection increases with an accessible primary forest. Men collect more than women
Mekonnen (1998)	Ethiopia	fuelwood dung	Resource access is important. On common lands: female youth are significant collectors, children contribute negatively. From household lands: men (and women?) are significant collectors.

Cooke (1998a)	Nepal's hills	fuelwood forage fodder	Evidence largely from commons. Collection time (CT): male and youth CT small, women's CT is large, women absorb the increase in CT due to increasing resource scarcity. Seasonal differences in CT: Men increase CT in off-peak agricultural season. Women increase CT (as joint product) when they are in the fields for seasonal agricultural activities. Youths increase their CT when women otherwise busy.
Cooke (1998)	Nepal's hills	fuelwood forage fodder	Evidence largely from commons. Increases in resource scarcity increase forest CT. Increases in CT do not decrease time for agricultural labor. Rather, collection increases during slack seasons (and increases come from forest production on own ag. lands?).
Kohlin (1998)	Orissa, India	fuelwood equivalents	Resource stocks from a) village woodlots (VWL, which are managed commons) and b) open access natural forests. Women, adolescents, higher caste collect from VWL. Lower caste collect from natural forest. Men collect more in general and MP(men) > MP(women), but VWL formation saves women more time.
Heltberg, Arndt, Sekhar (1998)	Rajasthan, India	fuelwood	Levels of resource stock and resource access significant. Some private tree substitution for open access stocks. Only women collect. Collection is seasonal, occurring when energy demands are high and demands for agricultural labor are low.

Cooke (1996, 1998) and Kohlin (1998) provide the most thorough inquiries into the question of labor allocation. Cooke focused on the seasonality of collection from the forest. Kohlin focused on labor allocation between alternative fuelwood sources. Cooke found that, in Nepal's hills, collection takes advantage of slack labor and opportunities for joint production. Women are the most important collectors of forest products. Nevertheless, men do collect and they increase their collection in off-peak seasons for agricultural labor. Women increase their collection in seasons when they spend more time away from the household itself and in the fields closer to the trees.[8] Youths increase their collection when adults, especially women, are otherwise occupied with peak season agricultural activities. And finally, the collection of forest products does not interfere with agricultural labor. Rather, overall increases in collection time tend to occur during slack seasons or to originate from reductions in time for leisure activities.

Kohlin found that, in Orissa, men actually collect more fuelwood than women do and that the marginal products of men for the collection activity are greater than the marginal products of women. Men, adolescents, and higher caste women do more of their collecting from village woodlots while lower caste women collect more from the less accessible natural forest—whether they are collecting for market supply, as we previously discussed, or for their own domestic use. Kohlin's households are sensitive to the source of their fuelwood. With increasing plantation stocks they tend to decrease both fuelwood collection from more distant natural forests and purchases of commercial fuel.

Thus, the story that began with our review of consumption patterns remains intact after our review of production patterns. That is, with sufficient scarcity, households do respond. On the consumption side, less-well-off households are more likely to feel the pressure of increasing scarcity and they respond by substituting alternative consumption goods and by increasing their reliance on their own productive abilities rather than on market purchases. On the production side, the literature suggests that household labor allocations are consistent with the economic rewards—rather than with external perceptions of absolute cultural norms like "women collect." Men may even collect more than women when the returns to male collection are greater than the returns to women's collection effort—although this may be unusual because other male wage opportunities are generally greater yet.

Some experience suggests that a) poor households respond to increasing scarcity by relying increasingly on more distant open access natural forests and some suggests that b) those households with land respond by growing the scarce product on their own private agricultural lands. The evidence is not thorough on these points and the identifying characteristics, particularly of the latter agroforestry households, is not clear. Our next section on adoption and investment addresses this issue.

Investments In New Technologies

Table 12.6 summarizes the six household econometric investment analyses known to us, plus one richly data intensive household survey (Scherr 1995) which is also presented from a perspective of multivariate impacts even if its method is not statistical. The specific parameters used in these analyses vary too widely for meaningful comparisons and none of them reported elasticities. Moreover, several of these analyses neglect factors of price and cost—despite the almost certain affect of these factors on a household's desire to invest. These studies provide useful insight nevertheless.

In all cases the empirical findings follow the expectations of economic behavior. Households invest when they anticipate gain in the form of higher prices or lower costs, and the most likely investors are households that can afford to take a chance on new investments of uncertain potential. These are households with larger incomes or greater wealth and households with "risk capital" in the form of larger incomes, or more labor or land than their neighbors, perhaps enough more to allow them to chance a small investment in an unproved new technology. The poorest households do not have the means to take this chance and the risk to them seems greater relative to the means they do have.

The investment literature tells us that better-off households invest first. Poorer households follow. The consumption literature told us that some of these poorer households are the more elastic respondents when they do invest.

The adoption of new consumption technologies is a more important question in urban areas where policy often encourages the substitution of commercial fuels as a means of decreasing the drain on forest resources. Since commercial fuels seldom penetrate rural markets, it is not surprising that the only analysis of rural adoption of a consumption technology features improved stoves. Amacher, Hyde, and Joshee (AHJ 1992) observed that wealthier households were the first to adopt improved stoves. They also observed that off-farm income is a more important predictor of adoption than income from a household's agricultural production. We might speculate that off-farm income indicates a greater variety of labor experiences and, therefore, broader household experience in general and greater exposure to information about new technologies.

Table 12.6. Characteristics of Households and Investments in New Technologies

Study	Location	Investment	forest prod. price	price var'n	Income farm/off-farm	sub-stitute	wealth or land owner-ship	land qual.	secur. of land tenure	h'hold labor total-male	train'g and ext.
Consumption Amacher, Hyde, Joshee (1992)	Nepal's hills	imprv'd stoves	+		insignif. +	-	+	NR	NR		
Production Amacher, Hyde, Rafiq (1993)	Pakistan	tree planting	+		+ +	-	+		+	+ +	
Mekonnen (1997)	Ethiopia	number of trees	NE		<----+---->	NE	+		+	+ +	+

Explanatory variables: sign of effect on investment

Study	Country	Dependent variable									
Thatcher, Lee Schelhas (1997)	Costa Rica	particpant: reforest pgm.	NE	-	+	NE	+	-	0	+	+
Patel, Pinkney, Jaeger (1995)	Kenya (Murang'a)	number of trees	NE	-	<----+---->	NE	+			+	
Scheer (1995)	Kenya (Siega and S. Nyanga)	number of trees	+	-	<----+---->	NE	+		+	+	
Shively (1998)	Philippines	number of trees	+	-	-	-	+				

Notes: a) NR: not relevant in the consumption form. NE: this critical variable was not estimated. b) AHR also noted the importance of the personal characteristics of extension agents. c) Mekonnen did not separate income sources. He identified a positive effect for increasing numbers of males in the household, both youths and adults, and an additional positive effect for male heads of household. d) TLS also observed a negative effect for improved land quality. Their measures of off-farm income and labor were i) off-farm income as a share of all income and ii) labor/hectare. e) PPJ used expenditure/capita as a proxy for all income, and restricted their measure of labor to labor available for farming. They observed that tree growing is more likely for higher wage households. They anticipate that higher wages mean higher incomes which suggest lower discount rates. The latter is an important factor for long-term investments like trees. f) Scheer did not observe prices directly, but she did observe more planting where returns are greater (prices higher or costs lower). g) Shively's observations were restricted to mango trees.

Investments in Trees

Two of the analyses of production-enhancing investments also identified the importance of access to information. In these cases (Amacher, Hyde, and Rafiq or AHR 1993; Thatcher, Lee, and Schelhas or TLS 1997), the new information was delivered through training programs or by extension foresters. In addition, AHR's Pakistani farmers took special notice of the personal characteristics of the extension agents who deliver the new information. The extension agent's good character determines the household's openness to information delivered by the agent. To these Pakistani farmers, the agent's character is even more important for successful technology transfer than the agent's forestry knowledge.

The literature on new production technologies raises additional points about the income, labor, and land variables, and introduces a risk spreading or income diversification argument that favors private investments in trees.

TLS offer a new perspective on income sources and labor availability. They observe that those Costa Rican households that are reliant on off-farm income and employment have less labor available for on-farm agricultural activities. These households may be more receptive to tree planting opportunities on their own lands because, unlike many agricultural crops, trees grow well with minimal labor input and that input can be scheduled for the slack seasons for off-farm employment. TLS addressed these propositions by reformulating the common income and labor variables as i) off-farm income as a share of all income and ii) labor available for farming.

Scherr examined differences in the labor input. She observed that while gender does make a difference, the relationship is complex. Investment decisions reflect distinctions in land tenure and off-farm employment and income as well as distinctions in gender—but it is difficult to separate the effects of these characteristics on the decision to invest in trees. For example, female heads of households in Scherr's regions of Kenya have less secure tenure and males are the greater participants in off-farm employment.

Most farm households have been used to collecting their forest products from the natural forest. For them, forest investments are a totally new experience—different in their use and timing of inputs and in their output markets, different from household experience with the natural forest and different from the improvements many households have grown to expect from new agricultural technologies.

Therefore, it is not surprising that Scherr observed that households invest only incrementally in new agroforestry technologies, beginning at a small scale and expanding gradually as the investment demonstrates its worth. Better-off households are generally the first to invest and Scherr observed that they are willing to invest in longer rotation species and for commercial timber production. When poorer households eventually invest in agroforestry, Scherr observed that they tend to plant short rotation fuelwood species. This is consistent with Patel, Pinkney, and Jaeger's (PPJ 1995) observation, for a different region of Kenya, that better-off farmers plant more. PPJ speculate that the better-off have (better access to credit and) lower time preferences.[9] Scherr speculates that a household labor constraint leads the poor to invest in fuelwood plantings as a means to save time spent collecting fuelwood.[10] In contrast, she speculates that land is the more constraining factor for investments in technologically advanced agroforestry practices like alley cropping that are more important for commercial production and more attractive to better-off households.

Risk and Uncertainty

Finally, first Dewees (1995), then Scherr and PPJ, observed the risk reducing characteristics of agroforestry investments in their cases in East Africa. Dewees and Scherr consider that the returns from investments in trees are less volatile than returns from agriculture. Scherr also observed attempts to diversify by planting a variety of tree species. PPJ took particular notice of the importance of tree planting as a factor in controlling environmental risks like erosion.[11]

Dewees' and Scherr's observations on the risk reducing characteristics of many East African investments in trees are related to Shively's (1998) observation that Philippine farmers plant more mango trees when the variation in mango prices is smaller. Scherr also showed that reductions in price variation can be as important to farmers as increases in product prices. All three observations on price variation are comparable to North American evidence that forest investments are a valid tool for diversifying investment portfolios. In North America, investments in trees yield returns that are less variable than average portfolio returns, as well as returns whose cycles run slightly counter to normal business cycles (See Binkley, Raper, and Washburn 1996). Is this a general pattern for forest investments in both developed and developing countries?

Overall Summation

The household literature explains a lot about the sequence in which different classes of farm households invest and why they invest, and it reinforces the consumption and production arguments to carefully select target populations for technology transfer and other development activities. Households invest i) because forest products are an important component of the household budgets of money or time, perhaps ii) for the income protection that diversification offers, and iii) because relative prices or imperfections in land and labor markets are sufficient to make this investment better than some alternative expenditure of household resources. The question that remains is "when are the relative forest product prices great enough?"[12] This question is especially important where forestry is an altogether new investment. Farmers who are entirely inexperienced with it will be the greatest benefactors from the advice offered in training sessions and from forest extension agents. When the relative prices are insufficient, however, no amount of additional information can create a good investment. Therefore, knowing when the prices are sufficient to justify investment in a new forest technology will predict when the various forms of technology transfer can be beneficial.

The consumption evidence suggests that prices are sufficient when we begin to observe substitution. We can anticipate that the same signal is appropriate for investments in new production technologies as long as we observe an important caution with respect to property rights for land and trees. Consider poor, and then better-off households, then consider community investment opportunities. Poor households may substitute other goods or alternate sources of the forest resource but they will not invest in new forest production technologies if their landholdings are insufficient even for more critical household products like food. Households with larger landholdings may substitute for scarce forest products but they may not invest in new production technologies when the period of land tenure is uncertain or where roads or government policies limit their access to markets. Communities may be slow to invest if they share rights to common lands with a central government agency and the system of shared rights is uncertain or confining. This final case is common because central governments retain legal ownership of most countries' forests and degraded rural lands, and because the central governments' rules for local uses of these lands are often restrictive, yet the central governments do not have the resources to enforce their rules completely and uniformly. We will revisit the question of private of community investment in the next section.

Unifying Principles

These observations generally follow a rather basic economic principle—resources (land, labor, and what limited capital these households possess) are allocated to higher valued uses until the marginal gains from their use equal their marginal costs. Indeed, the empirical and analytical literature seems to reject the strongest expected claim for a cultural constraint, *i.e.* women collect, in the presence of evidence that both men and women collect, and the question of "who collects" seems to depend on comparative economic advantages within the particular household.

Von Thunen's Pattern

A common pattern begins to emerge from the observations summarized in this review, a pattern of rural development, deforestation and increasing scarcity of forest products, and eventual substitution with forestry investments that could potentially limit further deforestation. This pattern is an application of the concepts of economic geography first proposed by von Thunen (Dickinson 1969). Our figure 12.1 captures the basic elements of this pattern. It also provides the key reference points necessary for further reflections on investment timing, institutional constraints, and the policies affecting social forestry.

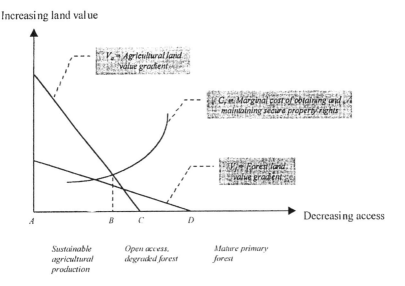

Figure 12.1. The Forest Landscape–A New Frontier

Figure 12.1 describes a simple landscape of agriculture and forests. Consider agricultural land first. The value of agricultural land is a function of the net farmgate price of agricultural products—which is greatest when the farmgate is near the local market at point A. Land value declines with decreasing access (which is closely related to increasing distance) as described by the function V_a. The function C_r describes the cost of establishing and maintaining secure rights to this land. This function increases as the level of public infrastructure and effective control declines and the cost of excluding trespassers increases with distance. Local communities may protect some lands beyond B to a declining degree further from their homes—as by sending children out to manage their grazing livestock—but eventually no number of forest guards can fully exclude illegal loggers and other open access users of remote forests.

The functions explaining agricultural land value and the cost of secure property rights intersect at point B. Farmers manage land between points A and B for permanent agricultural activities. They use land between points B and C (where agricultural land value declines to zero) as an open access resource to be exploited for short-term advantage. They harvest native crops that grow naturally in this region, crops like fodder for their animals, native fruits, and fugitive resources like wildlife. They do not invest even in modest improvements in the region between B and C because the costs of protecting their investments would be greater than the investments are worth. Their use of this open access region is unsustainable except in pulses of natural regrowth.

Initially, the mature natural forest at the frontier of agricultural development at point B has a negative value because it gets in the way of agricultural production and its removal is costly. Settlers remove trees whenever the agricultural value of converted forest plus the value of the trees in consumption exceeds their removal costs. Eventually, the most accessible forest resources will have been removed. Now the function V_f describing forest value must intersect the horizontal axis at some distance beyond B. Market demand for forest products may still justify their removal at this time, and it will continue to justify their removal as the forest frontier gradually shifts outward to some point like D. In this case the market price of forest products is just equal to the cost of their removal and delivery to the market. The *in situ* price at point D is zero, and the value of forestland at D is also zero. The region of unsustainable open access activities now extends from B out to either C or D, whichever is farther. The costs of obtaining and protecting the property rights insure that this region will remain an open access resource. Some governments protect some lands past point B but they must absorb the increasing protection costs—and even then trespass occurs. Some amount of illegal logging occurs almost everywhere in the world and no number of well-trained and well-motivated forest guards can prevent it. For example, some local citizens illegally harvest Christmas trees from the well-managed national forests in the eastern US. The US Forest Service does not extend great effort to prohibit this theft because the costs of enforcement would be greater than the potential gain.

The construct of figure 12.1 conforms to the common description of any initial settlement. In some cases, trees actually impede agricultural development, the forest rent gradient is very low, and point D can even be to the left of point C if net forest resource values are sufficiently low. Apparently, this describes the forest frontier in Cote d'Ivoire (Lopez 1998) and Bolivia's

Amazon region today (Bowles et al. 1997). In other cases, the region between B and D can be large (*e.g.*, in Nepal's hills or India's Rajastan) but the forest in this region is generally degraded. The positive net value of the original resource, together with the open access character of the region, has assured removal of the best resources. Some degraded vegetation remains in the region and it will re-grow naturally. The lowest wage households will continue to exploit these resources when the scattered vegetation grows to a minimum exploitable size or when its fruits begin to ripen (Amacher et al. 1993, Foster et al. 1997).

As the natural forest is depleted over time, the forest margin at D will gradually extend farther and farther from the market. Deforestation will continue, and the delivered costs of forest products will continue rising. The incentives are insufficient to induce tree planting and any attempt at forest management will be unsustainable. As Godoy (1992) points out, the prices of forest products may be rising, but they are not yet sufficient to induce forest management.

Eventually the margin at D extends far enough, and delivered costs and local prices become great enough, to induce substitution. Substitution may take the form of new consumption alternatives to forest products (for example, kerosene or improved stoves as substitutes for fuelwood), or it may be production related—as in planting and sustainable forest management on some land closer to the market. In Latin America alone there are 165 million hectares of this secondary forest, an area three-fourths the size of Mexico (Smith et al. 1998).

The forest rent gradient rises with the increase in delivered costs (following the arrows in figure 12.2) until it intersects the agriculture rent gradient to the left of agriculture's intersection with the property rights cost function. We might call this case a "mature" frontier. Forest product prices are now sufficient to justify the substitution of planted and managed forests for the resources of the open access natural forest. The new managed forests will occur in the region B'B" of figure 12.2. They may take the form of industrial timber plantations or they may take the forms of agroforestry or even of just a few trees planted around individual households. The latter are excluded from most measures of the forest stock but their economic importance can be large. In Bangladesh, for example, recall that they account for 3/4 of all marketed timber and fuelwood consumption (Douglas 1982). They are major sources of fuelwood consumption in Malawi (Hyde and Seve 1993), timber production in Kenya (Scherr 1995), and of positive environmental externalities in northern China (Yin and Hyde 1998) or the Murang's region of Kenya (Patel, Pinkney, and Jaeger 1995).

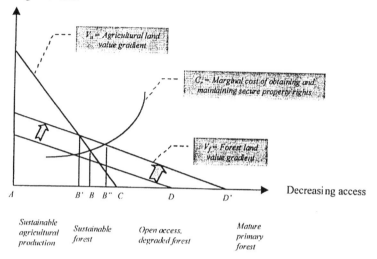

Figure 12.2. The Forest Landscape–A Mature Frontier

In all cases, removals from the mature natural stock are concentrated around a point like D in figure 12.1—or D' in regions characterized by the higher forest rent gradient in figure 12.2. Mature natural stocks in the region before D (or D') were removed in earlier times. In most cases a mature natural forest of no economic value exists beyond D (or D'). Sometimes this region is small (*e.g.*, in Ireland or Cape Verde). Sometimes this latter region continues well beyond D (Siberia, Alaska, northern Canada, much of the Amazon) until it can become much the largest share of reported physical stocks. Many countries possess regions of both cases because many forest products are either bulky or perishable and do not transport well. That is, standing natural reserves remain in some regions while the forests in other regions are depleted and some households in the latter may have begun planting on their own lands.

Evidence and Qualifications

A number of empirical studies have demonstrated this development pattern for commercial timber.[13] More recently, several have demonstrated its relevance for fuelwood and for agricultural expansion. In particular, Hofstad (1997) confirmed a pattern of expanding extraction for charcoal in the vicinity of Dar es Salaam and Chomitz and Griffiths (CG 1997) confirmed a similar pattern for charcoal from multiple population centers in Chad. CG also observed the substitution opportunities that constrain expanding supply regions and rising charcoal prices once a backstop energy source or technology price is attained. Their observation that substitution eventually constrains deforestation is consistent with observations from the household literature that substitution constrains the consumption of high cost fuelwood.

Most everyone is careful to recognize that market access on the horizontal axis of figure 12.1 is more complex than distance alone. It probably includes road quality and density and measures of both terrain and moisture: steep and swampy areas are less accessible than flat dry areas. Site quality can also be important for some land uses and Chomitz and Gray (1996), for example, were careful to incorporate an element of land quality in their assessment of agricultural expansion into the forested regions of Belize.

Most everyone would also recognize that there are many forest value gradients. Rather than the single sharply defined gradient V_f in our figures, each forest product has its own gradient. And most recognize that the cost function C_r varies with the individual and institutional particulars in the local market. It almost always includes an element of hazy uncertainty that it not fully acknowledged in our sharply drawn figures. (We will return to this latter point about property rights later.) Regardless of these adjustments, von Thunen's pattern prevails in general.

There is a data problem, however, that creates exceeding difficulty for accurate measures of the economic forest stock, for economic measures of resource scarcity, and for meaningful estimates of rates of deforestation. The common estimates of resource supply in forest policy analyses are actually measures of the standing physical stock. Adjustments for the difference between physical stock and economic supply are seldom incorporated in analysis and the potential for error is great. Alston, Libecap, and Mueller (1998) demonstrated the importance of infrastructure to accurate measures of secure economic supply. In particular, they demonstrated the negative relationship between access to population and government administrative centers and the costs of enforcing claims on land use in frontier regions of Brazil. (This is precisely the function C_r in our figures.) Hyde, Krutilla, Barnes, and Xu (1998) observed that the standard measures of physical resource stocks include a large component of forest beyond point D (or D'). They corrected for economic supply with adjustments for terrain, road density, and moisture. Their corrections explained forty percent of the variation in estimated physical stocks across 31 cities in eight tropical countries. General policy differences explained no more than twenty percent of the variation. The obvious point should be that elements of infrastructure and terrain are critical. They can dominate elements of policy, yet assessments of forest policy and forestland use only rarely account for them.

Further Development: The Rural Poor

So, the empirical evidence confirms the pattern of our von Thunen figures, but how do the figures and this evidence relate to the human populations and the production and consumption of those products of greatest concern for social forestry, the topic of our review? They argue that local users of the forest will continue to travel farther and farther to collect and deplete the natural forest until the local market prices or the implicit wage for their own collection time attains a level sufficient to induce substitution of a backstop technology. For households without land or without secure rights to their land, the backstop technology can be a consumption substitute like kerosene or improved stoves. For those with secure rights to their own land, it can be a production substitute like private or local community tree plantings. This is exactly what the household literature observed with respect to prices and wages or collection time, substitution, and adoption.

But what about the rural poor? They are the greatest concern of rural development, and several of the production and consumption assessments discussed earlier in this paper noted greater participation of the poorest households in the array of activities sometimes identified as "social forestry."

Foster, Rosenzweig, and Behrman (FRB 1997) addressed this question with a general equilibrium model of three sectors: agriculture, rural manufacturing, and open access forests. In their model, land is an input to agriculture and forests, but not to manufacturing, and forests are open access resources. FRB's model could be interpreted as similar to our figures with all manufacturing taking place at the origin. FRB's open access forest is, of course, the region to the right of B in figure 12.1 (or B" in figure 12.2). In their model, the marginal product of agricultural labor is its wage. The average product for forest labor falls until it equals its marginal product—in what is the standard observation for open access resources—and the wage for forest labor is below the wage for agricultural labor.

FRB confirmed their analysis with satellite data on land use and an extensive household survey from 250 villages in eleven states of India. One of their conclusions is that the lower wage for forest labor causes the removal of forest resources to extend past the optimum for forest properties with secure property rights—as our figures also show. Perhaps more interesting for us is the implication of their analysis that it is lower wage households and individuals who participate in extraction from the open access natural forest.

We would hypothesize that since individuals from lower wage households accept a lower return for their marginal effort they can afford to venture further into less accessible regions to obtain the products of the forest. And in the more accessible regions of depleted open access forests, individuals from lower wage households can afford to remove lower quality material that would not provide sufficient compensation for the effort of the better-off. The poor will renew their exploitation of the degraded open access regions as the depleted resources in these regions recover to minimal levels. These depleted resources will never attain a level sufficient to compensate the effort of the better-off. Of course, FRB's analysis and this hypothesis explain many observations around the world of the greater dependence of the poor and otherwise disadvantaged on the forest. Once more, they are consistent with observations from the household literature, observations that poorer households are the first to respond to increasing prices and they first respond by shifting from market purchase to household collection from the open access forest.

Finally, FRB showed that technical change has a varied effect on forest use. Technical change in the manufacturing sector raises wages and attracts labor from the other sectors. By drawing labor from the forest, it decreases dependence on the forest. Bluffstone (1995), with evidence from Nepal, anticipated that off-farm labor opportunities in general would have this effect. On the other hand, FRB showed that technical change in agriculture has a different effect. While technical change in agriculture also raises wages and attracts labor, it can deplete the forest by inducing land conversion to agriculture. We would anticipate that the new agricultural land would come from the region of point B for new frontier communities described by figure 12.1. It would come from the plantation forest region B'B" for the mature communities described by figure 12.2. Sunderlin and Wunder (1998) confirm these contrasting effects of manufacturing and agricultural sector growth on the forest with macroeconomic evidence from a broad sample of 67 countries.

Shortcomings in Data and Research

We have reviewed an extensive body of empirical analysis. Much of it is recent, and most of it is internally consistent. This literature does have its shortcomings, however, and it would be useful to identify them before drawing any conclusions for policy. The shortcomings fall into the general categories of data, the agriculture/forestry interface, and macroeconomic policy impacts. Each imposes cautions on the policy conclusions we draw in the final section of this paper and each reminds us of research questions that are yet unanswered.

The first set of shortcomings has to do with data. Many of the household analyses conducted to date feature South Asian examples. A few rely on examples from Asia outside of the Indian sub-continent and a few rely on African data. The particular dearth of Latin American examples may not be a serious shortcoming because the majority of Latin America is not a forest scarce region, and perhaps it is not a serious shortcoming if one accepts the general premise that human behavior is similar everywhere. In this case the important range of examples for social forestry would be the full spectrum of relative scarcities for the important local forest products regardless of their geographic identity. Nevertheless, it would add confidence to any generalizations we might draw if Latin American experiences were a greater part of the literature.

A second and potentially more important data shortcoming is that most household surveys shortchange the local landless population. This was not their intention but we did note in our discussion of household production that most surveys have more evidence of household market purchases of forest products than of household market sales. Since the sellers of forest products in subsistence economies tend to be lower income and landless households, this means many data sets under-represent these households. This is a particularly unfortunate shortcoming because the poorest households are the most important target of economic development.

A third data shortcoming has to do with reliance on physical rather than economic measures of forest stocks and, especially, physical measures that capture stocks of dense forest whether or not these stocks are economic while entirely disregarding the economic contribution of sparser groups of trees in fields, along hedgerows, in farmyards, and in degraded areas. Both corrections can be critical. We are reminded that HKBX's sample from 31 regions in eight tropical countries found that correcting for the error or uneconomic stocks accounted for three times the variation in established forest stocks due to policy differences. Meanwhile, we are also reminded that Douglas observed that 3/4 of market supply in Bangladesh originates from unmeasured, mostly private, stocks. Of course, the real problem is that the data- and analysis-intensive effort necessary to overcome these shortcomings would overwhelm most forest policy analyses.[14]

The second major analytical shortcoming largely has to do with the agriculture-forestry interface. As the natural stock of forest resources becomes scarcer, the economic reasoning and the empirical evidence pretty much establish that households eventually substitute land and labor away from activities like agricultural production and toward forest activities. Should we be concerned that these poor households may be foregoing some component of their agricultural subsistence? They may also switch labor away from household activities like nutrition and health care to forest activities. Or, stating it as a question "How much do human health and productivity suffer from declines in the forest cover?" In contrast, as development proceeds, improving wage labor opportunities may correlate with increased forest cover.

Most, not all, data sets and household analyses to date have restricted themselves to open access forest resources and to forest-related household activities. We cannot expect to understand the welfare impacts of social forestry without improving our understanding of the opportunities for alternate allocations of the household's most important productive inputs, its labor and its agricultural land.[15]

Finally, our questions have all been framed in the perspective of what scarcity of forest products means for substitution. We might also inquire what improvements in agricultural productivity have meant for the availability of land and labor for forest production. Does increasing agricultural productivity effectively, if indirectly, eliminate scarcity in forest products?

The third major shortcoming has to do with policy, specifically policy which causes either immigration or emigration to the rural and forested interior and alters the composition of the population of rural poor. Because neither immigrants nor emigrants are likely to be perfectly representative of the rest of the population, the population subsequent to imposition of the policy may have different consumption and production patterns. Our ability to predict the performance of the new population with respect to social forestry activities or anything else will suffer until we learn more about the policies that induce rural migration and the behavior of the populations that most commonly migrate. This is an important shortcoming because so many government policies and so much successful development is accompanied by migration either to or from the forested rural interior.

Final Policy Observations

Despite these research shortcomings, we have learned a tremendous amount about rural household responses to increasing scarcity of forest resources and about opportunities to introduce the technologies generally identified with social forestry. Despite forestry's common condition as a standing mature open access resource that is almost always available as a substitute for managed production, we have evidence that rural households follow systematic patterns of economic behavior in their production and consumption of forest resources. We also have evidence that managed and sustainable social forestry activities are successful in select economic situations within select regions and for select populations. Furthermore, their success often occurs among those poorest households who are the intended target of most development activity. We can identify these select situations. The remaining questions have to do with policy. Where should we look for policy opportunity? And what cautions does the evidence recommend for policy makers?

Our figures can help reflect on these two questions. All three functions in the figures can be affected by policy, and shifts in any of the three affect the management of either sustainably managed forests, or open access natural forests, or both. We will consider each in turn.

But first, let's consider government investments in infrastructure, in roads in particular. These investments affect the overall form of the von

Thunen model, improving access for all land uses and shifting the intersection points for the agriculture and forest land use gradients (C and D or D', respectively) to the right. The net effect on sustainable forest crops (region B'B" in figure 12.2) is uncertain, but road improvements also clearly extend the region of consumption from the open access natural forest. Therefore, road design is clearly important to local forest use and governments would be well-advised to design their road improvement programs to minimize access to forested areas that are critical for their non-market values.

Spillovers From Agricultural Policy

Agricultural incentives shift the agricultural rent gradient upward and outward. Where forestry is a sustainable activity and managed forests represented by the region B'B" in figure 12.2 do occur, then agricultural incentives make agriculture relatively more attractive than managed tree cover. They induce land conversion away from sustainable forestry and toward agriculture.

Agricultural incentives have a very different effect in regions that are still reliant on the open access natural forest. These regions are better described by figure 12.1. Increasing the agricultural rent gradient in this case induces land conversion as well, but from either the mature forest at new frontiers where forest removal is still a deterrent to agricultural expansion (the neighborhood of point C), or the degraded forest in longer established communities (between C and D). Agricultural incentives in cases characterized by figure 12.1 also raise the price threshold or backstop that must be overcome before forestry can become sustainable.

Therefore, we anticipate that agricultural incentives will always work against sustainable forestry activities, either currently or by delaying their application in the future. Revising ill-conceived agricultural incentives can only promote sustainable forestry, either now or once local forest product prices have risen sufficiently to justify sustainable activity.

Direct Forest Policies

Direct forest policies include incentives like cost sharing, free seedlings, technical assistance, investments in public lands, and public silvicultural research, as well as disincentives like taxes or restrictions on harvesting and shipment. Incentives raise the forest rent gradient and disincentives lower the gradient. The latter, especially disincentives on shipment, are common in developing countries. They include export bans and domestic processing requirements, any number of licensing requirements, and restrictions on the mills or regions within a country to which raw material shipment can be made.

Their objectives are generally to assist the local industry or the environment—although their effect is not always favorable for either one.

The impacts of either incentives or disincentives depend, once more, on whether the affected region is a region of sustainable forestry (described by figure 12.2) or a region that still relies on an open access natural forest (described by figure 12.1), and some forest policy instruments affect one region but not the other. For example, simply subsidizing the prices of forest products (or reducing taxes on them) raises the forest rent gradient and increases the area under sustainable management in figure 12.2. It also extends the area of open access forest depletion. Restrictions on shipments do the opposite. They decrease the number of potential competitive purchasers and, therefore, decrease the prices of forest products and shift the rent gradient downward, decreasing the area of sustainable management but also decreasing the area of open access forest depletion.

Seedling distribution and technical assistance, on the other hand, decrease forest management costs and raise the net value forest gradient in regions characterized by sustainable forestry. They have no effect on regions characterized by open access extraction because prices in these regions are insufficient to justify planting and households in these regions obtain no economic advantage from the free seedlings or the management advice offered by technical assistance.

These forest policy observations imply that siting can be an important aspect of social forestry. Kohlin's (1998) work with households in 23 villages in Orissa illustrates this point. (We discussed this work in our sections on consumption and production.) Kohlin observed that some villages gain from the introduction of new village woodlots while others do not. Location is an essential variable for the villages that do gain. The greater the distance between new woodlots and the natural forest, the greater the gain. Better-off households and individuals tend to use more accessible woodlots while the poor continue to rely on more distant natural forests.

Location is especially important in subsistence economies because the regional infrastructure tends to be poorly developed. Shipment of bulky and perishable products is restricted and markets are geographically limited. Since forest products are generally one or the other, bulky or perishable, a weak regional infrastructure can be particularly restrictive. Where the markets are all local, social forestry projects and target populations must also have a very local orientation. Geographically broader forestry projects risk incorporating both regions characterized by sustainable forestry and other regions characterized by newer frontiers and open access forest consumption—which may ensure project failure in one region or the other.

Fortunately, many social forestry development projects are local. Also fortunately, government extension programs provide an instrument that is capable of addressing local differences. Extension's role in technology transfer is particularly appropriate where forestry is a new "crop" with which local households have little prior management. If extension services can identify those communities where forest product prices are about to attain the threshold that justifies substitution, and if they can identify the appropriate technology for the conditions of land and labor availability that characterize the community, then their role as transfer agent of social forestry information can be vital for improving local welfare.

No empirical literature known to us fully describes how to predict when prices will reach the substitution threshold, or when some households and regions will be ready for the technology transfer. This is an unsolved problem worthy of real consideration. Once this problem is solved, the literature on adoption suggests that the poorest households are not the best initial targets for new technologies. Households with more resources, households that can afford to take some risk, are more likely initial adopters. The poor rapidly follow the leadership of successful adopters, however, and they may receive the greatest eventual benefit from many social forestry activities. Obtaining these benefits, however, is generally contingent on possessing secure land use rights for the properties on which households might plant and manage trees. This brings us to the final category of policy opportunities.

Secure Rights to Forest Properties

The third function in our figures reflects the costs of establishing and maintaining secure property rights. Decreasing these costs shifts the function downward and converts some of the degraded open access area to the right of B or B" in the figures into sustainable use—either for agriculture or for forestry. In regions where forestry is not yet sustainable, decreasing these costs reduces the price threshold that sustainable forestry must eventually overcome. Some promising policy options for social forestry; e.g., joint forest management and community forestry; can be described in terms of shifts in this third function.

The function captures two categories of activities, direct costs and uncertain rights. Direct costs are costs like title or contract registration and fencing. Uncertain rights refer to either the uncertainty the user has for use of the land or its forest resources, the uncertainty the user has about the overall policy environment. An uncertain policy environment affects long-term expectations for returns on investments in the land and, therefore, willingness to make long-term investments like forestry.

Direct Costs

The important prospects for reducing direct costs lie in improvements in the regional infrastructure that could make title registrations easier in formal economies. Simply opening local offices for registration or reducing the bureaucratic formality for registration are two examples. The counterpart to these in less formal economies is improved access to the village headman or other enforcer of local rules.[16] Of course, the right to easy transfer must accompany possession of secure rights. Otherwise, higher use values will eventually replace some historical values and create land use conflict and uncertainty even for those with formal rights to the land.

Formal rights can be uncertain even in the present. The classic case in forestry features a government ministry which holds the formal rights and which may attempt to enforce them, but which does not have the means to enforce full compliance with all of the ministry's rules for forest use. The many examples of this case have led to joint (ministry and local community) forest management (JFM) in India and to arguments for both privatization of natural forests and land transfers to community-based forestry in the Philippines, Indonesia, Colombia and numerous other countries. India's JFM now extends to seven million has. and 35,000 village forest protection commitees.[17]

JFM, Community Forestry, and Privatization

JFM provides some rights for local users but reserves other rights (usually for mature timber) for the government. Kant (1996) showed that two conditions are necessary for successful JFM: heavy local dependence on the forest and a population that is homogeneous in these demands. With a lower level of dependence, the local community has little incentive to invest its own resources in forest management and remaining open access stocks may be sufficient. Heterogenous demands suggest competing interests and difficulties in arriving at and respecting community decisions. Even where his two conditions are met, Kant observed that local demands on the forest vary across communities. Therefore, he argued that JFM contracts must be community specific. Furthermore, most of the household production literature shows that local demands are not homogenous. Women and men, landed and landless households, farmers and those with off-farm employment opportunities, the poor and the better-off all have different demands on the local forests. Therefore, communities characterized by both of Kant's conditions may be uncommon and successful applications of JFM may be very specific cases.

Finally, Johnson (1998) reminds us that sub-optimal rewards obtain when the residual claimant is not the party with greater ability to affect forest use. In both Johnson's example from Honduras and India's JFM, the government claims the residual timber but local users have greater impact on the forest. In Honduras, local resin tapping delays timber growth to maturity and the timber returns expected by the government.[18] This is an additional reason to anticipate that successful joint public agency and local community management operations may occur only infrequently.

Some recommendations for community forestry, and even private forest management, overcome these problems. They recommend either permanent transfers or long-term agreements with private contractors or local citizen groups—who have full rights rather than partial rights shared with the government agency, as in JFM. The argument for the transfer of rights to communities is that governments, as absentee landlords, have difficulty enforcing their rules or in taking advantage of locally specific land use opportunities. Local managers are better at both. Furthermore, if local groups managing the forest as a commons have a cost advantage over private individuals it is in protecting the forest for preferred common uses. (Their cost advantage is the equivalent of a reduction in the property rights cost function in our figures. The cost reduction extends the area of sustainable agriculture or plantation forest management and decreases the region of open access depleted natural forest.)

There are also important cautions for expectations of successful community forestry. First, the community management solution may not be permanent. As development proceeds, local relative values may be modified and some other lower cost management arrangement may replace the original community management. This is not a problem as long as the forest ministry expects and permits modifications in the contractual arrangements over time.

Second, some forest ministries impose serious restrictions on community forest activity. Each restriction only offsets the cost savings that transfers to local use were designed to create. Philippine plans for Community Based Forest Management (CBFM), for example, once required local hire of a professional forest manager, a forest inventory and plan, several steps of public and government agency oversight, restrictions on shipments of forest products beyond provincial boundaries, and payment of 44 percent of gross forest revenues to the ministry (Hyde et al. 1997). The Philippines' more recent success with CBFM has only begun after many of these restrictions were simplified or eliminated. Current discussions of transfers from the ministry to community management in Colombia, Nepal, and Zambia all include similar restrictions. We must recall that one reason for the transfers to community management is the ministry's prior inability to enforce rules like these. The ministries must minimize their restrictions on community management and focus their own efforts on enforcing simple rules for the remaining ministry objectives.

The Long-term Policy Environment

We initially identified a second category of uncertainty, an uncertain policy environment. Conflict in the contract between the formal owner (generally the ministry) and the community group with use rights is not the only source of uncertainty. Uncertainty can also arise from any number of external sources that raise doubts about continued land use. Consider a couple examples.

Deacon (1994), with data from 120 countries, observed that rates of deforestation increase in unsettled political environments; for example, in regions characterized by substantial political turmoil, military governments, or guerrilla activity. One explanation for Deacon's observations would be that, in unsettled political environments, those with rights to commercial forests perceive a risk of loosing their claims on these forest resources. Therefore, they claim the market reward to their rights while the rights are still in their possession. Those without rights may also feel that unsettled times provide them with greater opportunity for trespass and theft with impunity.

Yin and Newman (1997, updated and reconfirmed by Zhang, Uusivuori, and Kuuluvainen 1999), provided confirming evidence of short-term maximizing behavior for forest management under political uncertainty. They examined timber harvests and reforestation in two regions in China after individual farmers in communes were given long-term land use contracts. In one region the new contractual rights were administered consistently and the lands were fully redistributed within six years. In this case, farmers harvested timber and invested in reforestation. Authorities in the second region first distributed, then recalled, contracts for some forestlands. Despite a 15-year upward trend in land under long-term private contracts and despite rising timber prices, farmers who held contracts in this second region perceived the uncertainty in the local authorities' decision making, and they harvested without reforesting.

In our own experience, land managers in the southern Philippine island of Mindanao have been slow to invest in forestry: i) despite rapid growth in long-term capital investment in other sectors of the Philippine economy (indicating good macroeconomic conditions for forestry investment), and ii) despite evidence of forest plantations in neighboring Indonesia (indicating good biological and external market conditions for forest investment). We would hypothesize that the difference (until recently) has been a stable policy environment in both Indonesia and in other sectors of the Philippine economy but a Philippine policy of ever-changing environmental rules for the Philippine forest industry.

In each of these cases uncertainty made a risky prospect of land use and the rights to the standing forest resource or to new investments in forestry. The manager's uncertainty could be due to uncertainty in the broader economy, an unpredictable local political environment, or uncertainty specific to the prevailing rules for forestry.[19] In all cases, the source of the uncertainty is external to the behavior of the land manager, and in all cases it induces short-term unsustainable decisions characterized by resource depletion and deforestation without reinvestment. This behavior is identical to our expectations for behavior on open access forestland without any legitimate and enforceable claims. Removing the basis for the uncertainty effectively secures tenure for more land and forest resources and permits the land manager to obtain the rewards from investments in lands and forests. Therefore, removing the uncertainty lowers the cost curve for property rights and increases the likelihood of sustainable activities.

Conclusions

In conclusion, forestry, forest depletion, and the scarcity of fuelwood and other forest products consumed by rural households are not the problems many anticipated when Eric Eckholm (1976) called attention to these issues 25 years ago. We now have good empirical evidence that people adjust to changes in forest cover. Rural households all over the world display a large amount of flexibility in their consumption and production decisions for forest products and services.

Rural households eventually take advantage of opportunities to substitute for their consumption of forest products; for example, substituting agricultural residues and improved stoves for some fuelwood. In the event of real scarcity, however, their initial adjustments tend to occur on the production side, and household labor—or the time household's devote to collecting forest products—is as important as the physical scarcity of the forest resource. In particular, the opportunities households have to collect while jointly pursuing other household activities or to collect during seasons of slack labor use provide flexibility for many forest production activities.

As scarcity increases yet further, farm households do respond by planting trees. They treat trees as another crop and many of the experiences of subsistence agriculture become relevant for forestry. There are constraints on agroforestry activities, however. Land tenure and government policy can be critical deterrents. The poorest households often have no land to plant trees. And those households with land are reluctant to plant unless their tenure is certain for a period at least as long as the productive life of a tree crop. Where land and its tenure pose no problem, then government policy often acts as a deterrent to tree planting and management, sometimes creating an uncertain overall environment for investment of any sort, sometimes encouraging agriculture or some other land use in preference to forestry, and sometimes directly discouraging forestry itself by restricting the price or movement of forest products. Uncertain tenure and detrimental public policies reduce household expectations for forest product values, decrease the value of land in forest cover, and decrease the area of sustainable forestry.

We must recognize that even when the conditions of land tenure and government policy are entirely favorable, sustainable forestry will never be a universal phenomenon. The very best policy cannot make it so because the costs of establishing and maintaining secure property rights will always exceed the value of the standing forest resource on some frontier lands. Efficient household behavior will always treat lands described by this condition as open access resources and exploit the forests on these lands unsustainably. And for lands without pervasive positive externalities, efficient government policy will not alter this land allocation.

Nevertheless, the depletion of forest cover and the increasing relative prices of forest products that we observe in numerous regions around the world create an expanding opportunity for social foresty. Households and communities in a broad range of geographic and demographic situations have responded by planting and managing their own trees. The forest management responses of many farmers add up and the total impact on a country's forest cover can be great, despite the local limitations of many markets for forest products. Markets are often local because forest products are generally bulky and perishable and also because roads and other regional infrastructure are not well-developed.

If we accept this description of rural households and the rural environment, then foresters have a critical role. They can learn i) where the local relative prices are great enough, or almost great enough, to induce the introduction of social forestry technologies and ii) where the local experience with these technologies is limited. Foresters have a role as agents of technology transfer in these locations. The regions, target populations within a region, and even the sites and institutional arrangements for forestry assistance require careful selection for the foresters to be effective. Forest ministries and international donors can support this technology transfer role. They have two additional roles: i) identifying detrimental tenurial arrangements and unfavorable government policies and acting as advocates for their change, and ii) identifying and protecting areas of significant external social value within the regions of open access and unsustainable forest depletion. Altogether, the tree planting and management activities of rural households, technology transfer activities of foresters, and policy and protection activities of ministries and donors can improve the social welfare of rural households and increase the regions of sustainable land use. They must be selective and well-designed, however, because scarcity of forests and forest products is distinctly not a universal phenomenon.

NOTES

1. Centre for International Forestry Research, Bogor, Indonesia; Virginia Tech, Blacksburg, VA, USA; and Goteborg University, Goteborg, Sweden, respectively. Our chapter benefits from numerous presentations and from discussions with many who share our interests. We would like to thank Priscilla Cooke and Alemu Mekonnen especially.
2. Environmental accounting incorporates the value of non-market environmental and natural resource services in a country's system of national accounts. See Hyde and Amacher (1996) for a review of forest values in environmental accounts.
3. Dan Bromley and many others with connections to the University of Wisconsin were early leaders in the discussion of this issue. Jeff Romm and more recently Jeff Campbell of the Ford Foundation have encouraged the idea of formal local participation in what traditionally have been public or state forest properties. Many good scholars have developed the original themes of the Wisconsin and Ford insights.
4. Indeed, the initial conceptualizers of the dual economy were clear on this point. See Boeke (1948, 1953) and Furnival (1939, 1948), or see Ginsberg (1973) for a review. Boeke and Furnival featured households and populations that are predominantly subsistence oriented, but which always include at least occasional market participation. Neither Boeke's and Furnival's households nor their communities rely totally on subsistence production without any market exchange.
5. Wages appear in the demand functions of households whose production and consumption decisions are non-separable (Singh, Squire, and Strauss 1986). Some of the household forestry analyses calculate a shadow wage from the marginal product of labor in fuelwood collection.
6. AHJ's evidence contrasts with prior observations of many development practitioners that improved stoves and fuelwood consumption are correlates. In fact, stoves and consumption are correlates even with AHJ's data, but multi-variate techniques that separate the price, income, and improved stove effects on fuelwood consumption reveal stove substitution for fuelwood.
7. In contrast, HAS report that only women and children collect. It is not clear whether their survey allowed a male collection response. Collection did increase with the number of men in HAS's households in Rajasthan.
8. Mekonnen and AHJ reinforce this suggestion of joint production with another example. They observed that children spend time collecting forest products but their contribution to total collection is negative. This suggests that when women take children to the forest, they expect to produce something more than forest products. Perhaps instruction and child rearing are the joint products with collection from the natural forest. HAS added a third joint production possibility: collection while tending grazing livestock.
9. Sajise (1998), with evidence from the Philippines, is examining whether the income distributive incentives for new investments are more complex than yet recognized. The issues of income source and level will remain important in our review. We will revisit them yet again in our discussions of deforestation, property rights, and policy.

10. This observation is consistent with previous observations that fuelwood consumption and production are more wage than price responsive; in particular with AHK's observation that poor households in Nepal are the most wage responsive in fuelwood production.

11. Dewees and PPJ disagree in one important respect. Dewees believes farmers invest in trees because imperfect markets limit their other opportunities. Relying on evidence from the same region of Kenya, PPJ argue that once the erosion

12. Sajise is also examining the rate and timing of adoption, critical questions both for endangered environments and for estimates of the economic potential of long-term technology transfers. He points out that while the literature we have reviewed focuses on the decision of whether to invest, the investment timing decision can be even more important. That is, even if prices are high enough to justify investment, three characteristics of investments can explain landowner decisions to delay: i) their uncertain nature, ii) the sunk cost and, therefore, irreversible nature of investments in forestry, and iii) the fact that investments can always be postponed. The many examples of government policies that restrict private harvests or require licensing or other permits before transportation of privately produced forest products only enhance the second characteristic and reinforce the option to delay.

13. See Berck (1979) or Johnson and Libecap (1980) for examples from US history. Stone's (1998) assessment for Brazil's Paragominas is an especially clear example for a developing country and a modern context.

14. Chomitz' work is all the more important because its intensive use of GIS data does help overcome the problem (Chomitz and Gray 1996, Chomitz and Griffiths 1997).

15. Indeed, in some cases these relationships become very complex. For example, increasing the forest cover near stationary water can improve the habitat for mosquitos and increase the prevalence of malaria. On the other hand, one species, *Azadirachta indic,* or the neem tree, produces seeds and a chemical discharge that may be mosquito deterrents. Rural households in parts of East Africa and India plant this species near the household for that purpose.

16. For agriculture, Feder et al. (1988), with an example from Thailand, measured the productivity gains from title registration and showed that these are sufficient to cover the costs of the government titling activity. For sub-Saharan economies with a briefer formal history, Migot-Adholla et al. (1991) concluded that the informal system works well and there are few gains from establishing formal title.

17. Sushil Saigal, Ecotech Services, New Delhi; personal communication, January 28, 1999.

18. This is reminiscent of taungya, an agroforestry system that originated in Burma's hills a century ago. Taungya combines government agency objectives for a high-valued timber plantation crop with subsistence agricultural objectives. In exchange for protection of newly established forest plantations, the agency allows the local community the rights to produce agricultural crops on the plantation until the young tree growth shades out the agricultural crops (Nair 1991). Of course, taungya often fails because the community quickly perceives its own incentives to maintain the agricultural rights by deterring the forest growth.

19. An unstable macroeconomy would have the same effect because macroeconomic instability makes long-term rewards uncertain and, therefore, deters

long-term investments. We can observe an example in Indonesia since the beginning of the East Asian financial crisis in July 1997.

REFERENCES

Alston, L., G. Libecap, and B. Mueller. 1998. *Titles, Conflict, and Land Use: the Development of Property Rights and Land Reform on the Brazilian Frontier.* Ann Arbor: University of Michigan Press.

Amacher, G., W. Hyde, and B. Joshee. 1992. "The adoption of consumption technologies under uncertainty: a case of improved stoves." *Journal of Economic Development* 17(2):93–105.

Amacher, G., W. Hyde, and B. Joshee. 1993. "Joint production and consumption in traditional households: fuelwood and crop residues in two districts of Nepal." *Journal of Development Studies* 30(1):206–25.

Amacher, G., W. Hyde, and K. Kanel. 1998. "Nepali fuelwood production and consumption: regional and household distinctions, substitution, and successful intervention." *Journal of Development Studies* (forthcoming).

Amacher, G., W. Hyde, and M. Rafiq. 1993. "Local adoption of new forestry technologies: an example from Pakistan's Northwest Frontier Province." *World Development* 21(3):445–53.

Berck, P. 1979. "The economics of timber: a renewable resource in the long run." *Bell Journal of Economics* 10(2):447–62.

Bluffstone, R. 1995. "The effect of labor markets on deforestation in developing countries under open access: an example from rural Nepal." *Journal of Environmental Economics and Management* 29(1):42–63.

Boeke, J. 1948. *The Interests of the Voiceless Far East.* Leiden: Universitare Pers.

Boeke, J. 1953. *Economics and economic policy of dual societies.* New York: Insitute of Pacific Relations.

Bogahawatte, C. 1997. "Non-timber forest products and rural economy in conservation of wet zone forests in Sri Lanka." Unpublished report to Economy and Environment Program for Southeast Asia. Singapore.

Bowles, A., R. Rice, R. Mittermier, and G. Fonseca. 1998. Science 280:1899–1900.

Byron, N. 1997. "Income generation through community forestry." Unpubl. mss. Centre for International Forestry Research, Bogor, Indonesia.

Chomitz, K., and D. Gray. 1996. "Roads, land use, and: a spatial model applied to Belize." *World Bank Economic Review* 10(3):487–512.

Chomitz, K., and C. Griffiths. 1997. "An economic analysis of woodfuel management in the Sahel: the case of Chad." World Bank policy research working paper 1788.

Cooke, P., 1998. "The effect of environmental good scarcity on own-farm labor allocation: the case of agricultural households in rural Nepal." *Environment and Development Economics* 3(4):443–69.

Cooke, P. 1998a. "Intrahousehold labor allocation responses to environmental good scarcity: a case study from the hills of Nepal." *Economic Development and Cultural Change* 46:807–30.

Deacon, R. 1994. "Deforestation and the rule of law in a cross section of countries." *Land Economics* 70(4):414–40.
Dewees, P. 1989. "The woodfuel crises reconsidered: observations on the dynamics of abundance and scarcity." *World Development* 17:1159–72.
Dewees, P. 1995. "Trees on farms in Malawi: private investment, public policy, and farmer choice." *World Development* 23:1085–1102.
Dewees, P., and N. Saxena. 1995. "Wood products markets as incentives for farmer tree growing." In J. Arnold and P. Dewees (eds.). *Tree Management in Farmer Strategies*. Oxford: Oxford University Press.
Dickinson, H. 1969. "Von Thunen's economics." *Economic Journal* 79:894–902.
Douglas, J. J. 1982. "Consumption and supply of wood and bamboo in Bangladesh." Field document no. 2, UNDP/FAO project BGD/78/010, Bangladesh Planning Commission, Dhaka, Bangladesh.
Eckholm, E. 1976. *Losing Ground*. Washington: Worldwatch.
Feder, G., T. Onchan, Y. Chalamwong, and C. Hongladarom. 1988. *Land Policies and Farm Productivity in Thailand*. Baltimore: Johns Hopkins.
Foster, A., M. Rosenzweig, and J. Behrman. 1997. "Population and deforestation: management of village common land in India." Draft mss., Department of Economics, University of Pennsylvania.
Furnival, J. 1939. *Netherlands India: a Study of Rural Economy*. Cambridge: Cambridge University Press.
Furnival, J. 1948. *Colonial Policy and Practice: a Comparative Study of Burma and Netherlands India*. Cambridge: Cambridge University Press.
Ginsberg, N. 1973. "From colonialism to national development: geographical perspectives on patterns and policies." *Annals of the Association of Geographers* 63(1):1–21.
Godoy, R. 1992. "Determinants of smallholder commercial tree cultivation." *World Development* 20(5):713–25.
Heltberg, R., C. Arndt, and N. Sekhar. 1998. "Fuelwood consumption and forest degradation: a household model for domestic energy consumption in rural India." Draft mss., Institute of Economics, University of Copenhagen.
Hofstad, O. 1997. "Deforestation by charcoal supply to Dar es Salaam." *Journal of Environmental Economics and Management* 33(1):17–32.
Hyde, W., and G. Amacher. 1996. "Applications of environmental accounting and the new household economics: new technical economic issues with a common theme in forestry." *Forest Ecology and Management* 83(3):137–48.
Hyde, W., K. Krutilla, D. Barnes, and J. Xu. 1998. "The importance of economic corrections for physical stocks: a forestry example." Draft mss. Forestry Department, Virginia Tech.
Hyde, W., B. Harker, E. Guiang, and M. Dalmacio. 1997. "Forest charges and trusts: shared benefits with a clear definition of responsibilities." *Journal of Philippine Development* XXIV(2):223–56.
Hyde, W., and J. Seve. 1993. "The economic role of wood products in tropical deforestation: the severe experience of Malawi." *Forest Ecology and Management* 57(2):283–300.
IRG with Edgevale Associates. 1994. "The Philippine environment and natural resource accounting project (ENRAP phase II)." IRG, Inc. Washington.

Jodha, N. 2000. "Common property resources and the dynamics of rural poverty: field evidence from the dry regions of India." In W. Hyde and G. Amacher, *Economics of Forestry and Rural Development: An Empirical Introduction from Asia*. Ann Arbor: University of Michigan Press.

Johnson, R. 1988. "Multiple products, community forestry and contract design: the case of timber harvesting and resin tapping in Honduras." *Journal of Forest Economics* 4(2):127–45.

Johnson, R. and G. Libecap. 1980. "Efficient markets and Great Lakes timber: a conservation issue reexamined." *Explorations of Economic History* 17(3):372–85.

Kant, S. 1996. "The economic welfare of local communities and optimal resource regimes for sustainable forest management." Unpublished PhD thesis. University of Toronto.

Kapoor, R. 1992. "Investment in afforestation: trends and prospects." In A. Agarwal (ed.) *The Price of Forests*. New Delhi: Center for Science and Environment.

Kohlin, G. 1998. "The value of social forestry in Orissa, India." Unpublished PhD thesis. Economics Department, Goteborg University.

Kohlin, G. 1999. "Supply of fuelwood in rural India: evidence from Orissa." Draft manuscript. Economics Department, Goteborg University.

Lopez, R. 1998. "The tragedy of the commons in Cote d'Ivoire agriculture: empirical evidence and implications for evaluating trade policies." *World Bank Economic Review* 12(1):105–32.

Mekonnen, A. 1997. "Rural household fuel production and consumption in Ethiopia: a case study." Unpub. Ph.D. thesis, Economics Department, Gothenburg University.

Mekonnen, A. 1998. "Rural energy and afforestation: case studies from Ethiopia." Unpublished PhD thesis, Economics Department, Gothenburg University.

Migot-Adholla, S., P. Hazell, B. Barel, and F. Place. 1991. "Indigenous land rights systems in sub-Saharan Africa: a constraint on productivity?" *World Bank Economic Review* 5(1):155–75.

Nair, P. 1991. *An Introduction to Agroforestry*. Dordrecht: Kluwer.

Patel, S., T. Pinkney, and W. Jaeger. 1995. "Smallholder wood production and population pressure in East Africa: evidence of an environmental Kuznet's curve?" *Land Economics* 71(4):516–30.

Persson, R. 1998. "Fuelwood: crisis or balance." Unpublished draft mss, Centre for International Forestry Research, Bogor, Indonesia.

Rao, D. N., and M. K. Srivastava. 1992. "Role of forests in the energy, economy relationship." In A. Agarwal (ed.), *The Price of Forests*. New Delhi: Centre for Science and the Environment.

Sajise, A. 1998. "Information, labor constraints and the timing and intensity of agroforestry adoption." Draft manuscript. Agriculture and Resource Economics Department, University of California, Berkeley.

Scherr, S. 1995. "Economic factors in farmer adoption of agroforestry: patterns observed in western Kenya." *World Development* 23(5):787–804.

Schultz, T. 1964. *Transforming Traditional Agriculture*. Chicago: University of Chicago Press.

Sharma, R., M. McGregor, and J. Blyth. 1991. "The socioeconomic evaluation of social forestry in Orissa, India". *The International Tree Crop Journal* 7:41–56.
Shively, G. 1998. "Economic policies and the environment: tree planting on low income farms in the Philippines." *Environment and Development Economics* 3(1):83–104.
Shyamsunder, P., and R. Kramer. 1996. "Forest Farmers in the tropics: agricultural household model of forest resource use." Unpublished draft ms.
Singh, I., L. Squire, and J. Strauss. 1986. "The basic model: theory, empirical results, and policy conclusions." In I. Singh, L. Squire, and J. Strauss (eds.), *Agricultural Households Models*. Baltimore: Johns Hopkins University Press.
Smith, J, P. van de Kop, K. Reategui, I. Lombardi, C. Sabofal, and A. Diaz. 1998. "Dynamics of secondary forests in slash and burn farming: interactions among land use types in the Peruvian Amazon." Draft manuscript, Centre for International Forestry Research, Bogor, Indonesia.
Stone, D. 1998. "The timber industry along an aging frontier: the case of Paragominas (1990–1995)." *World Development* 26(3):443–48.
Sunderlin, W., and S. Wunder. 1998. "Mineral exports, Dutch disease, agriculture, and forests: new insights on the variability of tropical deforestation." Draft mss., Centre for International Forestry Research, Bogor, Indonesia.
Thatcher, T., D. Lee, and J. Schelhas. 1997. "Farmer participation in reforestation incentive programs in Costa Rica." *Agroforestry Systems* 35:269–89.
Yin, R., and W. Hyde. 1998. "Trees as an agriculture sustaining resource." *Agroforestry Systems* (forthcoming).
Yin, R., and D. Newman. 1997. "Rural reforms: the case of the Chinese forestry sector." *Environment and Development Economics* 2(3):289–304.
Zhang, Y., J. Uusivuori, and J. Kuuluvainen. 1999. "Impacts of economic reforms on rural forestry in China." Draft manuscript, Forestry Faculty, University of Helsinki.